Series Editor
John M. Walker
School of Life and Medical Sciences
University of Hertfordshire
Hatfield, Hertfordshire, AL10 9AB, UK

For further volumes:
http://www.springer.com/series/7651

Metabolic Profiling

Methods and Protocols

Edited by

Georgios A. Theodoridis

Department of Chemistry, Aristotle University of Thessaloniki, Thessaloniki, Greece

Helen G. Gika

School of Medicine, Aristotle University of Thessaloniki, Thessaloniki, Greece

Ian D. Wilson

Department of Surgery and Cancer, Imperial College London, London, UK

Humana Press

Editors
Georgios A. Theodoridis
Department of Chemistry
Aristotle University of Thessaloniki
Thessaloniki, Greece

Helen G. Gika
School of Medicine
Aristotle University of Thessaloniki
Thessaloniki, Greece

Ian D. Wilson
Department of Surgery and Cancer
Imperial College London
London, UK

ISSN 1064-3745 ISSN 1940-6029 (electronic)
Methods in Molecular Biology
ISBN 978-1-4939-9252-2 ISBN 978-1-4939-7643-0 (eBook)
https://doi.org/10.1007/978-1-4939-7643-0

Preface

This book provides a number of protocols for "global metabolic profiling," also known as metabonomics and/or metabolomics. Metabolomics deals with the holistic analysis of small molecules aiming to characterize the metabolic content of the studied samples/systems and reveal changes that result from alterations to them as a result of, e.g., different physiological states or the onset and progression of disease, etc. Over the last few decades, there have been significant developments in both analytical technologies and multivariate statistical methods that have greatly facilitated the growth of these holistic analytical approaches.

In putting together this volume, the editors have placed emphasis on obtaining chapters that illustrate the different approaches taken by researchers to develop tools to address the important challenges of the field. The first part of the book contains chapters on the challenges and perspective of the topic (Gika et al.), the use of quality control measures (QC) and validation issues (Begou et al.), data mining (Riccadonna and Francheschi), and bio- and chemoinformatic tools for metabolomics (Witting). These chapters highlight basic concepts such as experimental design, data treatment, metabolite identification, the need for harmonization, and the linking of data obtained by different analytical modes (also combining metabolomics results with data from other omics fields).

The second section, which is concerned with methodology, describes protocols for sample preparation centered on techniques for tissues, feces, and blood samples (Michopoulos, Deda et al., also addressed by Vorkas et al.) and chemical derivatization for GC-MS (Hušek et al.). The methods used for metabolite analysis and profiling are covered with chapters on GC-MS metabolic profiling (Klapa et al.), LC-MS profiling using both targeted methods (Virgiliou et al.) and IPC-LC-MS (Michopoulos), and untargeted (Want) profiling approaches. The profiling of polar charged metabolites still remains a challenge, and this section includes a chapter on the use of CE-MS for this purpose (Ramautar). NMR spectroscopic methods for profiling biological fluids (Benaki and Mikros) are also considered.

The volume concludes with two application sections covering the use of metabolomics in life sciences with examples of methodologies that can be found in food science or biomarker discovery for disease diagnosis and human well-being. In the case of food and natural products, the protocols describe the analytical methods used and their application in food quality control, where the use of NMR spectroscopy is described (Schripsema and Dagnino) and the evaluation of product authenticity and geographical origin (Spyros and coworkers). Both these issues represent major challenges for the food industry and are still a great concern for the health of the consumer. The use of proton-transfer-reaction time-of-flight mass spectrometry (PTR-TOF-MS) for the analysis of volatile organic chemicals (VOCs) is also described (Farneti). Arapitsas and Mattivi describe a protocol on the analysis of wine by LC-MS with application to the classification of wine according to the grape variety. In the case of applications in life sciences, the use of metabolic profiling for biomarker discovery in cardiovascular disease (Vorkas et al.) and the targeted analysis of steroids (Rudaz and coworkers) are described. Finally, Siopi and Mougios discuss experimental design and considerations on sample collection for studies involving human subjects.

While still an area of rapid technical development, the place of "omic" metabolic phenotyping where the objective is to gain unbiased, global knowledge of the content of the studied system, is firmly fixed as a means of gaining insights into the conditions under study, thereby enhancing our knowledge and detailed understanding of the phenomena under investigation.

Thessaloniki, Greece
Thessaloniki, Greece
London, UK

Georgios A. Theodoridis
Helen G. Gika
Ian D. Wilson

Contents

PART III PLANT/FOOD APPLICATIONS

PART IV LIFE SCIENCE APPLICATIONS

Contributors

M.R. ABELLONA U • *Section of Biomolecular Medicine, Division of Computational and Systems Medicine, Department of Surgery and Cancer, Faculty of Medicine, Imperial College London, London, UK*

MARIA AMARGIANITAKI • *NMR Laboratory, Chemistry Department, University of Crete, Heraklion, Crete, Greece*

PANAGIOTIS ARAPITSAS • *Department of Food Quality and Nutrition, Research and Innovation Centre, Fondazione Edmund Mach (FEM), San Michele all'Adige, Italy*

OLGA BEGOU • *Department of Chemistry, Aristotle University of Thessaloniki, Thessaloniki, Greece*

DIMITRA BENAKI • *Department of Pharmaceutical Chemistry, National and Kapodistrian University of Athens, Athens, Greece*

JULIEN BOCCARD • *School of Pharmaceutical Sciences, University of Geneva, University of Lausanne, Geneva 4, Switzerland; Swiss Center of Applied Human Toxicology (SCAHT), University of Basel, Basel, Switzerland*

DENISE DAGNINO • *Grupo Metabolômica, Universidade Estadual do Norte Fluminense, Campos dos Goytacazes, Rio de Janeiro, Brazil*

PHOTIS DAIS • *NMR Laboratory, Chemistry Department, University of Crete, Heraklion, Crete, Greece*

OLGA DEDA • *Department of Chemistry, Aristotle University of Thessaloniki, Thessaloniki, Greece*

BRIAN FARNETI • *Genomics and Biology of Fruit Crop Department, Research and Innovation Centre, Fondazione Edmund Mach, San Michele all'Adige, Italy*

PIETRO FRANCESCHI • *Computational Biology Unit, Research and Innovation Centre, Fondazione E. Mach, Trento, Italy*

HELEN G. GIKA • *School of Medicine, Aristotle University of Thessaloniki, Thessaloniki, Greece*

DAGMAR HANZLÍKOVÁ • *Institute of Laboratory Diagnostics, Department of Biochemistry, University Hospital Ostrava, Ostrava, Czech Republic*

PETR HUŠEK • *Institute of Laboratory Diagnostics, Department of Biochemistry, University Hospital Ostrava, Ostrava, Czech Republic; Biology Centre, Institute of Entomology, Analytical Biochemistry & Metabolomics, Czech Academy of Sciences, České Budějovice, Czech Republic*

IVA KARLÍNOVÁ • *Biology Centre, Institute of Entomology, Analytical Biochemistry & Metabolomics, Czech Academy of Sciences, České Budějovice, Czech Republic*

MARIA I. KLAPA • *Metabolic Engineering and Systems Biology Laboratory, Institute of Chemical Engineering Sciences, Foundation for Research & Technology - Hellas (FORTH/ICE-HT), Patras, Greece; Department of Chemical and Biomolecular Engineering, University of Maryland, College Park, MD, USA; Department of Bioengineering, University of Maryland, College Park, MD, USA*

ALEXANDRA KOLESNIKOVA • *NMR Laboratory, Chemistry Department, University of Crete, Heraklion, Crete, Greece*

TIIA KUURANNE • *Swiss Laboratory for Doping Analyses, University Center of Legal Medicine Geneva and Lausanne, Centre Hospitalier Universitaire Vaudois and University of Lausanne, Lausanne, Switzerland*

JIA V. LI • *Section of Biomolecular Medicine, Division of Computational and Systems Medicine, Department of Surgery and Cancer, Faculty of Medicine, Imperial College London, London, UK; Centre for Digestive and Gut Health, Institute of Global Health Innovation, Imperial College London, London, UK*

CHRISTONIKI MAGA-NTEVE • *Metabolic Engineering and Systems Biology Laboratory, Institute of Chemical Engineering Sciences, Foundation for Research & Technology - Hellas (FORTH/ICE-HT), Patras, Greece; School of Medicine, University of Patras, Patras, Greece*

EFI MANOLOPOULOU • *NMR Laboratory, Chemistry Department, University of Crete, Heraklion, Crete, Greece*

GEORGIOS MARKAKIS • *NMR Laboratory, Chemistry Department, University of Crete, Heraklion, Crete, Greece*

FULVIO MATTIVI • *Department of Food Quality and Nutrition, Research and Innovation Centre, Fondazione Edmund Mach (FEM), San Michele all'Adige, Italy; Center Agriculture Food Environment, University of Trento, San Michele all'Adige, Italy*

FILIPPOS MICHOPOULOS • *IMED Oncology, AstraZeneca, Macclesfield, Cheshire, UK*

EMMANUEL MIKROS • *Department of Pharmaceutical Chemistry, National and Kapodistrian University of Athens, Athens, Greece*

MARIA MISIAK • *NMR Laboratory, Chemistry Department, University of Crete, Heraklion, Crete, Greece*

VASSILIS MOUGIOS • *School of Physical Education and Sport Science at Thessaloniki, Aristotle University of Thessaloniki, Thessaloniki, Greece*

RAUL NICOLI • *Swiss Laboratory for Doping Analyses, University Center of Legal Medicine Geneva and Lausanne, Centre Hospitalier Universitaire Vaudois and University of Lausanne, Lausanne, Switzerland*

MATTHAIOS-EMMANOUIL P. PAPADIMITROPOULOS • *Metabolic Engineering and Systems Biology Laboratory, Institute of Chemical Engineering Sciences, Foundation for Research & Technology - Hellas (FORTH/ICE-HT), Patras, Greece; Division of Genetics, Cell & Developmental Biology, Department of Biology, University of Patras, Patras, Greece*

FEDERICO PONZETTO • *Swiss Laboratory for Doping Analyses, University Center of Legal Medicine Geneva and Lausanne, Centre Hospitalier Universitaire Vaudois and University of Lausanne, Lausanne, Switzerland*

EVANGELIA RALLI • *NMR Laboratory, Chemistry Department, University of Crete, Heraklion, Crete, Greece*

RAWI RAMAUTAR • *Leiden Academic Center for Drug Research, Leiden University, Leiden, The Netherlands*

SAMANTHA RICCADONNA • *Computational Biology Unit, Research and Innovation Centre, Fondazione E. Mach, Trento, Italy*

LUCIE ŘIMNÁČOVÁ • *Biology Centre, Institute of Entomology, Analytical Biochemistry & Metabolomics, Czech Academy of Sciences, České Budějovice, Czech Republic*

SERGE RUDAZ • *School of Pharmaceutical Sciences, University of Geneva, University of Lausanne, Geneva 4, Switzerland; Swiss Center of Applied Human Toxicology (SCAHT), University of Basel, Basel, Switzerland*

MARTIAL SAUGY • *Center of Research and Expertise in Anti-Doping Sciences, University of Lausanne, Lausanne, Switzerland*

JAN SCHRIPSEMA • *Grupo Metabolômica, Universidade Estadual do Norte Fluminense, Campos dos Goytacazes, RJ, Brazil*

PETR ŠIMEK • *Biology Centre, Institute of Entomology, Analytical Biochemistry & Metabolomics, Czech Academy of Sciences, České Budějovice, Czech Republic*

AIKATERINA SIOPI • *School of Physical Education and Sport Science at Thessaloniki, Aristotle University of Thessaloniki, Thessaloniki, Greece; Department of Physical Education and Sport Science at Thermi, Aristotle University of Thessaloniki, Thessaloniki, Greece*

APOSTOLOS SPYROS • *NMR Laboratory, Chemistry Department, University of Crete, Heraklion, Crete, Greece*

ZDENĚK ŠVAGERA • *Institute of Laboratory Diagnostics, Department of Biochemistry, University Hospital Ostrava, Ostrava, Czech Republic*

SOFIA TACHTALIDOU • *NMR Laboratory, Chemistry Department, University of Crete, Heraklion, Crete, Greece*

GEORGIOS A. THEODORIDIS • *Laboratory of Forensic Medicine and Toxicology, Department of Chemistry, Aristotle University of Thessaloniki, Thessaloniki, Greece*

CATHERINE G. VASILOPOULOU • *Metabolic Engineering and Systems Biology Laboratory, Institute of Chemical Engineering Sciences, Foundation for Research & Technology - Hellas (FORTH/ICE-HT), Patras, Greece; Human and Animal Physiology Laboratory, Department of Biology, University of Patras, Patras, Greece*

CHRISTINA VIRGILIOU • *Laboratory of Forensic Medicine and Toxicology, Department of Chemistry, Aristotle University of Thessaloniki, Thessaloniki, Greece*

PANAGIOTIS A. VORKAS • *Section of Biomolecular Medicine, Division of Computational and Systems Medicine, Department of Surgery and Cancer, Faculty of Medicine, Imperial College London, London, UK*

ELIZABETH J. WANT • *Computational and Systems Medicine, Imperial College London, London, UK*

IAN D. WILSON • *Department of Surgery and Cancer, Imperial College London, London, UK*

MICHAEL WITTING • *Research Unit Analytical BioGeoChemistry, Helmholtz Zentrum München – German Research Center for Environmental Health, Neuherberg, Germany; Chair of Analytical Analytical Food Chemistry, Wissenschaftszentrum Weihenstephan für Ernährung, Landnutzung und Umwelt, Technische Universität München, Freising, Germany*

HELENA ZAHRADNÍČKOVÁ • *Biology Centre, Institute of Entomology, Analytical Biochemistry & Metabolomics, Czech Academy of Sciences, České Budějovice, Czech Republic*

Part I

Fundamentals

Chapter 1

Metabolic Profiling: Status, Challenges, and Perspective

Helen G. Gika, Georgios A. Theodoridis, and Ian D. Wilson

Abstract

Metabolic profiling has advanced greatly in the past decade and evolved from the status of a research topic of a small number of highly specialized laboratories to the status of a major field applied by several hundreds of laboratories, numerous national centers, and core facilities. The present chapter provides our view on the status of the remaining challenges and a perspective of this fascinating research area.

Key words Metabolomics, Metabonomics, Biomarker, Metabolite identification, MetID, Biochemical pathway

1 Introduction

The field of untargeted metabolic profiling, also known as metabonomics/metabolomics [1, 2] or metabotyping [3], involves the study of the small molecule complement of samples such as biological fluids (plasma, serum, urine) cells, organs, or whole organisms. The earliest examples of the use of "holistic," untargeted, and hypothesis-free metabolic phenotyping can perhaps be traced back to the work of Dent and Dalgliesh [4, 5] who, in the late 1940s, used two-dimensional paper chromatography for the discovery of new disease biomarkers. This early work was followed by studies by Pauling and colleagues based on the use of gas chromatography to profile urinary volatiles for disease diagnosis [6–8]. Pioneering work on metabolic fingerprinting based on the use of liquid chromatography was also being undertaken at this time (e.g., [9, 10]).

It can be argued that such studies comfortably precede genomic (and proteomic) profiling by some years and falsify the repeated, and erroneous, statement that metabolic phenotyping is the latest addition to the "omics field," rather than the first! However, it was only with the availability of analytical systems that combined rapid, multianalyte detection with an element of structural information that, when used in combination with multivariate statistical analysis,

Georgios A. Theodoridis et al. (eds.), *Metabolic Profiling: Methods and Protocols*, Methods in Molecular Biology, vol. 1738, https://doi.org/10.1007/978-1-4939-7643-0_1, © Springer Science+Business Media, LLC, part of Springer Nature 2018

enabled the relatively rapid detection and identification of the potential biomarkers hidden in the profiles. These techniques, centered on ^{1}H nuclear magnetic resonance (NMR) spectroscopy and mass spectrometry (MS), with the latter often hyphenated to a separation such as liquid or gas chromatography (LC, GC), permitted the full development of the field and resulted in a rapid increase in the number of research groups active in metabolic phenotyping. This increase in research activity was accompanied by a dramatic, and still increasing, rise in the volume and sophistication of publications on the topic (*see* Fig. 1) and, while still lagging somewhat in number behind those of the genomics and proteomics, has now reached many hundreds per year.

However, with this expanding application of metabolic phenotyping in virtually all areas of the life sciences, there has been an increasing realization of the need for careful study design and standardization of methodology. This has also been accompanied by a better understanding of the advantages and limitations of the "holistic" approach to biomarker discovery and to the rise of more "targeted" approaches. So, untargeted methods have the major advantage that they are unencumbered by the preconceptions of the investigator and are therefore more likely to discover novel, and unexpected, metabolites. They are therefore ideally hypothesis-free, but hypothesis generating. However, a problem with this approach can be that many metabolites are seen to change in the test population relative to the control group that are not specific to the condition under investigation but rather represent a general response to effects such as stress or environmental factors. Disentangling these general, nonspecific changes from the direct effects on metabolites more directly/mechanistically involved in the process(es) under investigation can sometimes require significant effort.

Another related concern is how comprehensive the coverage of the metabolome can be using the techniques available. This again is by no means a trivial question as, unlike, e.g., the human genome which has been fully sequenced and the number of genes estimated, it cannot yet be claimed that the extent of the metabolome has been fully mapped. Also, the range of concentrations that the metabolites encompass from, presumably, zeptomoles to micromoles makes demands on the available analytical techniques that are, to say the least, challenging. While there have been very significant, and continuing, advances in analytical technologies, these have, in some ways, only served to highlight their limitations. Indeed, while the minimum specification for the "perfect metabotyper" is easy to produce, with such instruments required to be universally sensitive, nonspecific, unbiased, rapid, robust, and stable, quantitatively possessed of a large dynamic range and providing enough information on each of the components of the sample

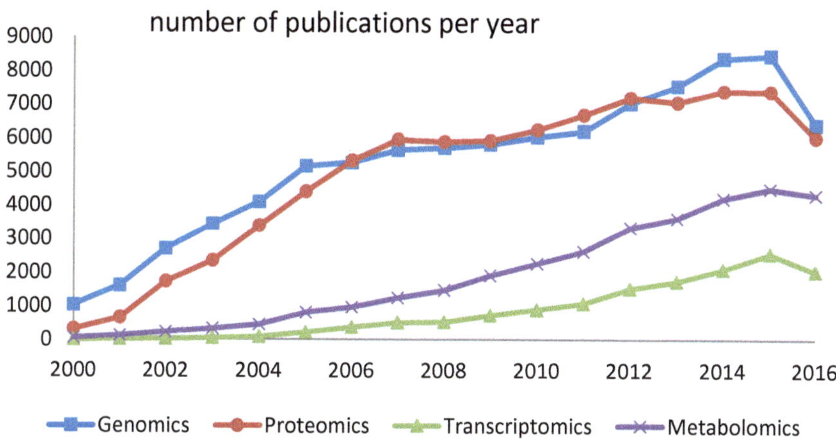

Fig. 1 Publication trends in the areas of genomics/transcriptomics, proteomics, and metabolic phenotyping from 2000 to 2016 generated from references contained within SCOPUS using the search terms genomics, proteomics, metabolomics or metabonomics or metabolic profiling, and transcriptomics (search made June 2017)

as to provide unequivocal identification, comparison with what is currently available for this type of work reveals a large gap in capability. And indeed, although several technologies have some of the characteristics of the hypothetical "perfect metabotyper," there is no single analytical solution/platform that provides the means of obtaining comprehensive metabolic profiles.

For this reason attempts at ensuring the maximum recovery of metabolite information currently depend on the use of several analytical platforms and methods. In the case of ^{1}H NMR spectroscopy, the use of 600 MHz instruments provides a good compromise between field strength, sensitivity, dynamic range, and affordability. Such instruments offer a means for the rapid analysis of biofluid samples such as urine and plasma and provide informative spectral data permitting structural characterization and identification (helped by the availability of good spectral libraries) [11]. In terms of the ease of use for the analysis of liquid samples, ^{1}H NMR spectroscopy usually requires minimal sample preparation other than filtering and pH adjustment. Solid, or semisolid samples such as tissues, can be analyzed either following extraction into a suitable solvent or in the native state using "magic angle spinning" NMR spectroscopy (although uptake of this technology in metabolic profiling applications has been limited) [11]. A typical example of the information available from the use of ^{1}H NMR spectroscopy in a comparative toxicology study on acetaminophen (APAP) and its related, less toxic, meta-isomer AMAP is shown in Fig. 2 for "aqueous" liver extracts obtained from control and treated mice [12]. These spectra can readily be used to distinguish between extracts obtained from different treatments and show the presence of both

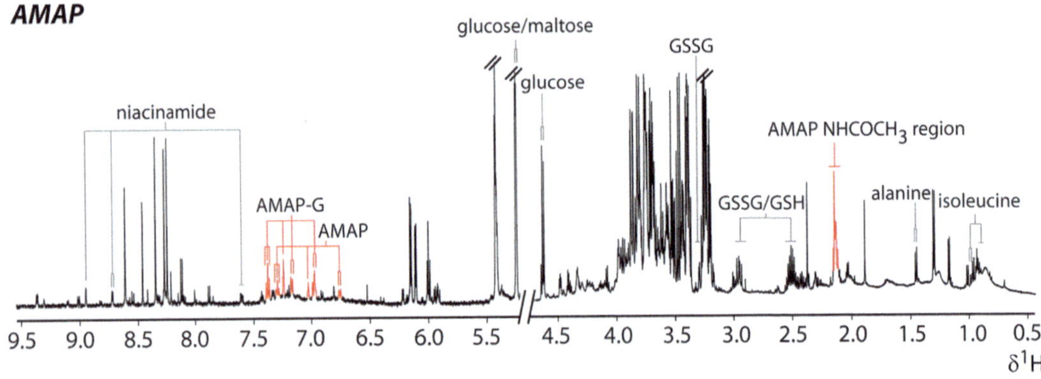

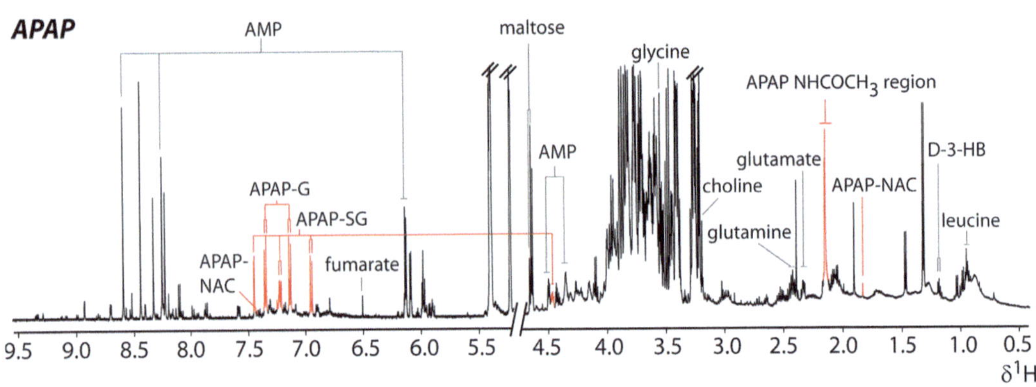

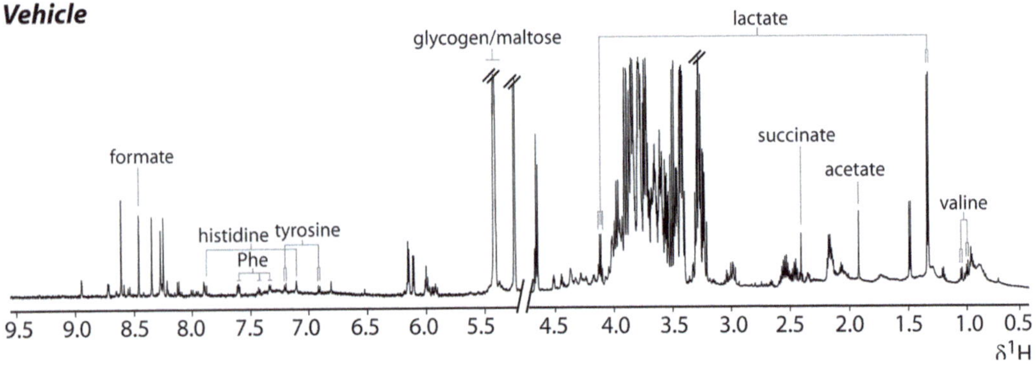

Fig. 2 Representative ^{1}H-NMR spectra of hepatic extract metabolic profiles of the acetaminophen (APAP), AMAP, and control groups at 1 h. Resonances assigned to drug-related molecules have been colored in red. Key: *APAP/AMAP-G* APAP/AMAP glucuronide, *APAP-SG* APAP glutathionyl, *APAP-NAC* APAP-*N*-acetylcysteinyl, *APAP/AMAPNHCOCH₃* APAP/AMAP *N*-acetyl resonance, *GSH* reduced glutathione, *GSSG* oxidized glutathione, *Phe* phenylalanine, *d-3-HB* d-3-hydroxybutyrate, *AMP* adenosine monophosphate, overlapped resonances from glucose/glycogen/maltose labeled. From reference [12] reprinted with permission

drug metabolites and endogenous metabolites present in the samples and give an insight into the metabolic changes that result from the administration of the test compounds to this species. A particular advantage of NMR spectroscopy-based methods is that they are inherently reproducible, and, e.g., one-dimensional ^{1}H NMR spectra acquired, at the same field strength on the same sample, should give very similar results irrespective of the laboratory or instrument manufacturer.

Mass spectrometry can, like NMR spectroscopy, be used for the direct analysis of liquid (or gaseous) samples by the simple expedient of directly infusing the samples into the ion source of the instruments ("DIMS") [13], and this approach, using either flow injection analysis or specialist interfaces such as the "nanomate," has many advantages in terms of simplicity and speed. However, particularly in the case of complex matrices such as urine- or blood-derived samples, ion suppression can be problematic. In addition it can be difficult to determine isobaric or isomeric compounds if they are present as mixtures using this approach. This has led to the adoption of the use of hyphenated techniques involving a chromatographic or electrophoretic separation prior to MS. Thus liquid and gas chromatography-mass spectrometry (LC-MS, GC-MS) and capillary zone electrophoresis-MS (CZE-MS or CE-MS) are very widely used for metabolic phenotyping, with LC-MS-based techniques currently in the ascendancy. GC-MS-based methods are, of course, particularly well suited to the analysis of volatile analytes present in, e.g., breath [14]. However, as will be evident from even a brief survey of the metabolic phenotyping literature, GC-MS-based methods have found a widespread use in metabolic phenotyping for involatile metabolites following chemical modification, via carefully optimized "derivatization" protocols, to make the analytes volatile [15]. This derivatization step should be studied meticulously since different metabolites react with different rates and the reaction conditions applied can affect the outcome of the analysis [16]. Procedures are available for the preparation of volatile derivatives of most analyte classes such as amino acids, sugars, polar acids, etc., and these can be applied to suitable extracts for a broad range of sample types (plasma, urine, plant extracts, food, etc.). Clearly, the need for fairly extensive sample preparation and derivatization required for the analysis of involatile metabolites means that getting to the point of actually performing the GC-MS analysis can be quite time-consuming and labor intensive. However, GC-MS with electron impact (EI) is a mature and robust technology, supported with a numerous databases of spectral data (NIST, Fiehn, etc.) to support the identification of potential biomarkers [14].

In the case of LC-MS-based metabolic phenotyping, the current state of the art employs methods based on so-called ultra (high) performance LC separations (UPLC, UHPLC) based on the use of high operating pressures and stationary phases formed from sub 2 μm-sized particles [17–19]. Chromatography on such phases provides excellent chromatographic efficiency enabling high-resolution separations to be obtained in reasonable analysis times (typically 5–15 min). Chromatography is performed using solvent gradient-based reversed-phase (RP) separations on C-18-bonded (or similar) phases on columns of 5–15 cm in length and flow rates from 200 to 900 μL/min. An example of the type of data that can be acquired using gradient RP-UPLC-MS is shown in Fig. 3 [19]. As this figure shows, while a large number of compounds elute in the first few minutes of the analysis, by careful optimization of the solvent gradient used metabolites can be spread throughout the whole of the analysis time. Sample preparation for liquid samples such as urine can often be limited to dilution and centrifugation to remove particulates. For protein-containing samples such as serum or plasma, it is first necessary to remove the proteins as these would otherwise irretrievably damage the column, but this is usually easily achieved via protein precipitation, typically via the addition of organic solvents. These reversed-phase separations have been shown to be suitable for medium polar to nonpolar analytes but are less good for polar ionic species, and for these types of analytes, the use of an alternative such as hydrophobic interaction liquid chromatography (HILIC) provides a partial

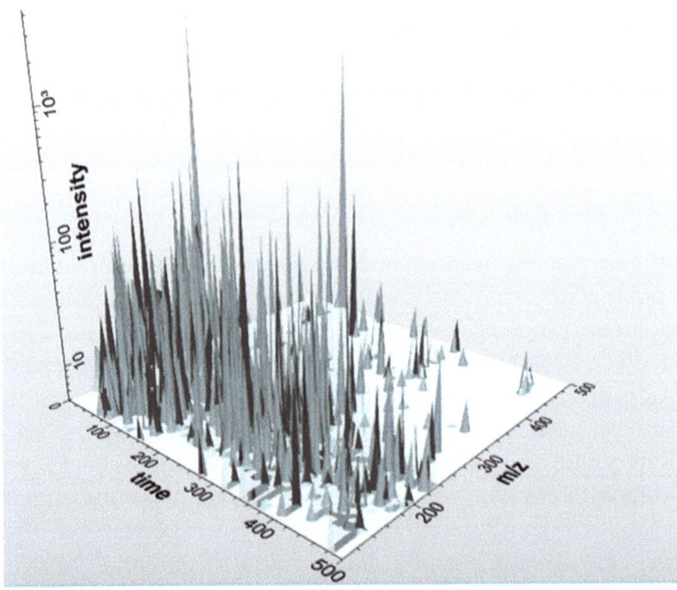

Fig. 3 Representation of a 3D mass chromatogram obtained from the reversed-phase analysis of rat urine on UPLC-TOF-MS. Reprinted from reference [19] with permission

solution being suitable for some, but not all, such compounds [19, 20]. For polar/ionic compounds that are unsuitable for analysis by HILIC-MS, the only remaining option may be to use ion pair (IP) LC where a suitable charged molecule, e.g., tributyl ammonium, is added to the mobile phase as an oppositely charged counter ion to "pair" with the oppositely charged analytes [21]. But, while effective, the use of IPLC generally requires the (effectively) permanent dedication of the system to this mode of operation thereafter, as decontaminating the instrumentation to remove all traces of the IP-reagent can be challenging. An alternative to LC-based methods for polar ionic compounds is, of course, to employ a capillary electrophoresis for the separation, and CE-MS methods have indeed shown utility in this role [22].

The upshot of all of this is that, in order to obtain the most comprehensive metabolite profile of a sample set possible, it may require more than one chromatographic system and analysis in both positive and negative modes of ionization (generally electrospray ionization (ESI), but also possibly APCI).

Having developed a suitable separation, a number of challenges remain in order to ensure that the analytical data that are obtained are useful. Unlike NMR spectroscopy-based methods, which are generally very robust, those utilizing a hyphenated MS have a number of challenges that need to be addressed. These result from the tendency of the analytical system (column and detector) to become modified over the course of the analysis. This can lead to minor changes in retention time, sensitivity, and (less often) mass accuracy. The existence of such effects requires the use of careful quality control procedures that can be used to monitor the analysis and correct for analytical drift of whatever sort. Various methods have been proposed for ensuring the validity of the data, of which one of the most common is the use of so-called quality control or QC samples. These are most often generated by making a representative bulk pool sample from aliquots of the samples to be analyzed. Typically it is first necessary to equilibrate the LC system by the repeated injection of a number of QC samples, which results in stable retention times. After this the QC sample is injected at regular intervals throughout the sample analysis [23]. After the run is completed, the data from these QC samples can be analyzed using multivariate statistical procedures such as principal component analysis (PCA), which provides a powerful tool to reveal trends in data that would indicate time-related (or other) effects that compromise the outcome. Assuming that the data passes such preliminary scrutiny, further measures to optimize it, e.g., peak alignment, can be performed, and the data can be examined for the presence of potential biomarkers. This part of the process relies heavily on the correct choice and the correct function of software tools. Some software still operate as black boxes, not providing much information or offering much freedom in the selection of parameters. This kind of software also performs different levels of multivariate statistics, such as PCA, and offers options for advanced visualization plots. Recently the applicability of open-source software (including web-based data treatment servers) has increased to a great extent. Such tools necessitate basic-level knowledge of programming and use of software language in R or MATLAB environment, but they offer unparalleled freedom in optimizing and tailoring the data treatment and data scrutiny process. They also offer advanced control in the visualization of the findings and the generation of plots, tables, and illustrations. Such tools can provide impressive outputs; however, statistical analysis is not yet proof and the famous Benjamin Disraeli quote is still timely [24]. Hence researchers are advised to pay exceptional attention to verify the validity of their findings by the use of different statistical analysis tools. Researchers who are new to the field are advised to resort to the assistance of fellow researchers who are more experienced to the specific topic of metabolome statistical analysis. Indeed the concept and the needs are different from those applied in statistics for genomics. Hence the statistical analysis tools to be selected and/or their fine-tuning for effective data treatment in metabolomics may vary to a great extent from the treatment of genomic data.

Having found peaks that appear to be correlated with the condition being studied, the next task is identification. This is often by no means trivial. For techniques such as NMR spectroscopy and GC-MS, there are large databases of spectral information that can be interrogated that may provide clues or even positive identifications of the metabolites of interest. In the case of LC-MS, the current situation is less promising as reliable databases for ESI-based spectra are still under development. Even when it seems possible to identify target metabolites from such databases, it can often be difficult to obtain authentic standards to confirm these tentative identifications. In such circumstances the complete characterization of the unknowns using a range of MS techniques is required, and modern instruments can enable high-resolution MS data to be obtained, including scan analyses such as MS-MS/MSn experiments in data-dependent acquisition mode, MS/MS with different collision energies (e.g., MSE), and so forth. By combining different levels of data, e.g., accurate mass with isotope ratio measurement from both precursor and product ions to obtain molecular formulae, and, e.g., comparison of the predicted fragmentation patterns for the tentatively identified metabolite with those obtained by experiment, it may be possible to increase confidence in the putative identification of unknowns. In the end, however, it may be necessary to isolate and identify unknowns using complementary methods, such as NMR spectroscopy, or alternatively synthesize standards when feasible.

Having identified potential biomarkers, there are a number of further considerations that require addressing. The first of these is biological context and plausibility—how likely is it that these molecules are involved mechanistically in the phenomenon under investigation? Directly? Or are they merely changes as part of a "global" response by the system as it tries to maintain homeostasis, etc. Indeed biochemical pathway analysis remains another key challenge in the development and maturation of omics-based biomarker discovery. The last few years have seen increasing efforts being invested in the development and application of software for pathway analysis. These software, either commercial- or Internet-based free web servers, accept lists of compound names or tables with metabolite concentrations, and their outcome lists indicate the most affected biochemical pathways. Such tools provide different levels of sophistication of their visualization tools. The overall aim is to utilize well-documented meta-analysis tools and data from public databases such as the Human Metabolome Database (HMDB) and/or KEGG (Kyoto Encyclopedia of Genes and Genomes) in order to offer tools able to systematically highlight the metabolites (and their corresponding pathways) with the most important perturbations in the studied data set [25].

Having obtained some level of confidence that there is a direct link to the system under investigation, there is a need to accurately quantify these putative biomarkers (and possibly related molecules in the same pathway) using targeted, validated assays. The use of such methods to reanalyze the sample set can confirm not only that the targeted molecules have indeed responded in the way seen in the untargeted assay but can also provide quantitative concentration data, rather than mere "fold change" results. Having achieved this level, the next step should be to analyze samples from other studies to confirm that the findings of the initial investigation are indeed valid.

This whole path is not an easy endeavor; however, it is in our view a necessary process that the research community needs to undertake to promote scientific knowledge in various aspects of the human activity, from farming and agriculture to environmental, medical, and the life sciences. This is because the metabolome directly reflects the current status of the biological system under investigation. Another reason is the fact that changes in the metabolome are the final result of the gene and protein function and as such are expressed (multiplied) to a large scale in relation to, e.g., the single nucleotide polymorphisms which caused their perturbation. Yet another reason is that (e.g.) in certain food products such as olive oil, wine, or honey, there is hardly anything else to look for: the content of such a sample is massively dominated by small metabolites. For these and many other reasons, we believe that the field of metabolic profiling will continue to increase. When the technological issues are finally overcome, we could expect stronger growth, rapid maturity, and larger application in routine analysis and operation. In the meantime additional research efforts and investments are necessary.

References

1. Nicholson JK, Lindon JC, Holmes E (1999) "Metabonomics": understanding the metabolic responses of living systems to pathophysiological stimuli via multivariate statistical analysis of biological NMR spectroscopic data. Xenobiotica 29:1181–1189

2. Fiehn O, Kopka J, Dörmann P et al (2000) Metabolite profiling for plant functional genomics. Nat Biotechnol 18:1157–1161

3. Gavaghan CL, Holmes E, Lenz E et al (2000) An NMR-based metabonomic approach to investigate the biochemical consequences of genetic strain differences: application to the C57BL10J and Alpk:ApfCD mouse. FEBS Lett 484:169–174

4. Dent CE (1952) Lectures on the scientific basis of medicine, vol 2. Athlone Press, London

5. Dalgliesh CE (1956) Two-dimensional paper chromatography of urinary indoles and related substances. Biochem J 64:481–485

6. Teranishi R, Mon TR, Robinson AB et al (1972) Gas chromatography of volatiles from breath and urine. Anal Chem 44:18–20

7. Pauling L, Robinson AB, Teranishi R et al (1971) Quantitative analysis of urine vapor and breath by gas-liquid partition chromatography. Proc Natl Acad Sci U S A 68:2374–2376

8. Robinson AB, Pauling L (1974) Techniques of orthomolecular diagnosis. Clin Chem 20:961–965

9. Scott CD, Chilcote DD, Lee NE (1972) Coupled anion and cation-exchange chromatography of complex biochemical mixtures. Anal Chem 44:85–89

10. Scott CD, Chilcote DD, Katz S et al (1973) Advances in the application of high resolution liquid chromatography to the separation of complex biological mixtures. J Chromatogr Sci 11:96–100

11. Lenz EM, Wilson ID (2007) Analytical strategies in metabonomics J. Proteome Res 6:443–458

12. Kyriakides M, Maitre L, Stamper BD et al (2016) Comparative metabonomic analysis of hepatotoxicity induced by acetaminophen and its less toxic meta-isomer. Arch Toxicol 90:3073–3085. https://doi.org/10.1007/s00204-015-1655-x

13. Han J, Danell RM, Patel JR et al (2008) Towards high-throughput metabolomics using ultrahigh-field Fourier transform ion cyclotron resonance mass spectrometry. Metabolomics 4:128–140

14. Theodoridis G, Gika HG, Wilson ID (2011) Mass spectrometry-based holistic analytical approaches for metabolite profiling in systems biology studies. Mass Spectrom Rev 30:884–906. https://doi.org/10.1002/mas.20306

15. Kopka J (2006) Current challenges and developments in GC–MS based metabolite profiling technology. J Biotechnol 124:312–322

16. Moros G, Chatziioannou AC, Gika HG et al (2017) Investigation of the derivatization conditions for GC–MS metabolomics of biological samples. Bioanalysis 9:53–65

17. Gika HG, Theodoridis GA, Plumb RS et al (2014) Current practice of liquid chromatography–mass spectrometry in metabolomics and metabonomics. J Pharm Biomed Anal 87:12–25

18. Rainville PD, Theodoridis G, Plumb RS et al (2014) Advances in liquid chromatography coupled to mass spectrometry for metabolic phenotyping. TrAC Trends Anal Chem 61:181–191. https://doi.org/10.1016/j.trac.2014.06.005

19. Theodoridis GA, Gika HG, Want EJ et al (2012) Liquid chromatography–mass spectrometry based global metabolite profiling: a review. Anal Chim Acta 711:7–16

20. Gika HG, Wilson ID, Theodoridis GA (2014) LC–MS-based holistic metabolic profiling. Problems, limitations, advantages, and future perspectives. J Chromatogr B Analyt Technol Biomed Life Sci 966:1–6

21. Michopoulos F, Whalley N, Theodoridis G et al (2014) Targeted profiling of polar intracellular metabolites using ion-pair-high performance liquid chromatography and-ultra high performance liquid chromatography coupled to tandem mass spectrometry: applications to

serum, urine and tissue extracts. J Chromatogr A 1349:60–68

22. Ramautar R, Nevedomskaya E, Mayboroda OA et al (2011) Metabolic profiling of human urine by CE-MS using a positively charged capillary coating and comparison with UPLC-MS. Mol BioSyst 7:194–199. https://doi.org/10.1039/c0mb00032a

23. Gika HG, Theodoridis GA, Wingate JE et al (2007) Within-day reproducibility of an HPLC–MS-based method for metabonomic analysis: application to human urine. J Proteome Res 6:3291–3303

24. https://en.wikipedia.org/wiki/Lies,_damned_lies,_and_statistics

25. Xia J, Wishart DS (2010) MetPA: a web-based metabolomics tool for pathway analysis and visualization. Bioinformatics 26:2342–2344. https://doi.org/10.1093/bioinformatics/btq418

Chapter 2

Quality Control and Validation Issues in LC-MS Metabolomics

Olga Begou, Helen G. Gika, Georgios A. Theodoridis, and Ian D. Wilson

Abstract

Global metabolic profiling (untargeted metabolomics) of different and complex biological matrices aims to implement an holistic, hypothesis-free analysis of (potentially) all the metabolites present in the analyzed sample. However, such an approach, although it has been the focus of great interest over the past few years, still faces many limitations and challenges, particularly with regard to the validation and the quality of the obtained results. The present protocol describes a quality control (QC) procedure for monitoring the precision of the analytical process involving untargeted metabolic phenotyping of urine and plasma/ serum. The described/suggested methodology can be applied to different biological matrices, such as biological biofluids, cell, and tissue extracts.

Key words Quality control, Untargeted metabolomics, Biological samples

---

1 Introduction

Metabolomics or metabonomics, two terms interwoven with each other, represent an expanding research discipline, dealing with the holistic analysis of metabolites (small molecules with molecular weight typically lower than 1500 Da). Basically, holistic profiling aims to provide a snapshot of the metabolic phenotype (metabotype) and to monitor changes of the endogenous profile of living systems in response to biological stimuli or genetic manipulation [1].

The field of metabolomics exhibited significant development in the last decade, especially due to the advancement of new technology platforms such as mass spectrometry (MS) and NMR (mainly H^1 NMR) spectroscopy [2]. In combination with each other, as well as with other technologies, e.g., liquid and gas chromatography (LC and GC) and advanced, sophisticated, multivariate statistical tools, these analytical techniques allow the simultaneous measurement of hundreds of endogenous compounds in different matrices, such as blood, urine, cells and various

Georgios A. Theodoridis et al. (eds.), *Metabolic Profiling: Methods and Protocols*, Methods in Molecular Biology, vol. 1738, https://doi.org/10.1007/978-1-4939-7643-0_2, © Springer Science+Business Media, LLC, part of Springer Nature 2018

tissues [3, 4]. However, the comprehensive analysis, simultaneous monitoring and relative quantification of numerous compounds face challenges due to the different physicochemical properties, chemical classes, and concentration ranges of the metabolites present and the complexity of the biological matrix [5–8].

Metabolomic studies can follow two distinct paths as either untargeted or targeted approaches. Targeted methods (including semi-targeted methods) are often hypothesis-driven analyses and as such can be focused on the measurement of specific metabolites or specific metabolic pathways (as described in more detail in Chapters 5 and 6 of the present book). In targeted analyses, Quality Control can be applied, with method evaluation criteria and the analytical figures of merit such as repeatability, analytical accuracy, and precision (for a review on targeted metabolomics, *see* ref. 9). Validation of targeted methods, although not trivial, is easier to perform, compared to untargeted methods, because predefined metabolites are analyzed and standards/reference solutions can be used. In contrast, untargeted metabolomics, by its very nature, focuses on the unbiased analysis of the sample of interest with the aim of the discovery of uncharacterized/unexpected biomarkers [10]. In such a case, validation and quality assurance of the analysis is much harder, especially when mass spectrometry is used coupled to a chromatographic technique, where factors such as retention time and MS-detection sensitivity may change during the course of a run [4, 11]. Hyphenated techniques, such as LC and GC-MS, inevitably generate high-dimensional data requiring extensive and complex data processing [12]. As a result, and despite its wide applicability, only a small number of articles have reported validation strategies for global metabolic profiling (e.g., *see* ref. 13).

A tested partial solution to this problem is the use of "quality control" (QC) samples whose role is to evaluate the stability and precision of the analysis [2, 14–16]. The use of QC samples provides information about the analytical performance of the overall system and also strengthens the analyst's confidence in the quality of the acquired data. Usually, QC samples are used for analytical platform equilibration, for monitoring the analytical signal allowing for intra-/inter-day precision evaluation, for signal correction (normalization), and method standardization [2, 15, 16]. QC data can be used as quantitative indicators of random errors or fluctuations during the analytical run. As a result, analysis of QC samples in metabolomic studies has a greater value than just the evaluation of chromatographic and mass spectrometry performance.

The idea for the QC samples is to prepare them by mixing small and equal aliquots from the real samples of interest so that the resulting pool sample (QC) will contain a mean concentration of all the metabolites present in the real samples. Such an approach may function for small- to medium-scale studies of up to ca 200 samples. For larger-scale studies, such schemes may not be possible; for practical reasons, a bulk sample of the same matrix may be substituted [17, 18], while sometimes both a bulk matrix and the study samples are used to provide a double QC, with the former used to enable analyses to be compared over long time periods/different batches [19]. The utility of this concept, or variations of it, has gained recognition over the past few years and has been widely applied for the analysis of endogenous compounds in a variety of different matrices, such as urine, plasma, serum, cells, and tissues [11, 15, 16, 20–22].

In the present protocol, we describe the use of QC samples during the analytical procedure involved in the reversed-phase LC-MS-based global metabolic profiling of biological samples. With adaption, similar procedures can be employed for other types of chromatography.

2   Materials

All solvents (methanol, acetonitrile, formic acid) used should be of LC-MS analytical grade. Water should be of Millipore quality (18.0 MΩ, at 25 °C). Standards are of analytical or higher grade. All standards should be stored at −20 °C or −80 °C (*see* **Note 1**).

2.1 Stock, Working, and Calibration Solutions

Stock solutions of all metabolites should be prepared at a concentration of 1 mg/mL (or appropriate concentration depending on analyte solubility) in methanol or mixtures of methanol with water. Working standards at a concentration of 10 μg/mL are prepared from stock solution by dilution with ultrapure water. From the stock solutions of the standards, prepare a mixture (concentration 1–5 μg/mL) of the compounds of interest by mixing appropriate volumes.

All solutions should be stored at −20 °C.

2.2 Mobile Phase

– Mobile Phase A: Water + 0.1 vol.% formic acid: add 1 mL of formic acid to 1 L of Millipore water.

– Mobile Phase B: Acetonitrile + 0.1 vol.% formic acid: add 100 μL of formic acid to 1 L of LC-MS grade acetonitrile. In the case of serum samples use methanol instead of acetonitrile or mixture of methanol with acetonitrile as mobile phase B.

– Wash Solvent: Use a "strong" solvent acetonitrile/water 80:20 v/v for post-injection cleaning cycles and a weak solvent water/acetonitrile 80:20 v/v for pre-injection washes (*see* **Note 2**).

2.3 Chromato-graphic Materials and Instrumentation

Chromatographic analysis can be performed on a HSS T3 C18 column (Waters, 2.1 mm × 150 mm, <2 μm particle size) or similar. Pre-columns filters and guard columns of the same material as the analytical column should be also used.

An ultrahigh performance liquid chromatography (U(H)PLC) system coupled to a high-resolution mass spectrometer with an ESI source can be used.

2.4 Software

Appropriate software include but are not limited to the following:

Data acquisition and processing software (e.g., Excalibur, MassLYnx, Analyst, or other). Special software for peak picking such as Marker Lynx, Sieve, MarkerView (or other vendor software), or open-source/free software (XCMS, MzMine, MetAlign or other). Microsoft Excel, Statistica, and other advanced spreadsheet software. SIMCA-P or other software for multivariate statistical analysis. MATLAB software and programming language or associated software packages: The programming language R is a popular and easy solution for data analysis, statistical computing, and graphics support.

3 Methods

3.1 Analytical System Preparation

1. Before starting, ensure that the mass spectrometer is in a suitable condition for the analysis of the samples. Check the mass accuracy of the mass spectrometer; if necessary, calibrate the mass spectrometer following the appropriate procedure recommended by the vendor to achieve maximum mass accuracy and resolution. It is advisable that the calibration procedure should take place every 3–4 months and that the temperature of the laboratory should not vary significantly.

2. Load the required liquid chromatographic method by setting parameters such as flow rate, column and autosampler temperature, wash cycles, and gradient elution program.

3. Allow the system to equilibrate and run a no-injection gradient in order to aid column equilibration. Carefully examine the results for this blank run for evidence of system/solvent contamination etc., and decontaminate if necessary.

4. Depending on the matrix of the samples being analyzed, run a suitable number of QC samples (prepared as described below) to achieve system stability. It has been observed that for urine analysis, 5 QC replicates are needed and for serum or plasma 10–20 injections (*see* **Notes 3** and **4**).

3.2 Sample Handling

1. It is advisable to always divide all samples on collection into an appropriate number of sub-aliquots for later storage and also in order to avoid unnecessary freeze/thaw cycles.

2. If possible, store sub-aliquots in different freezers, in case a freezer malfunction takes place. Ideally, samples should be stored in freezers as soon as possible after sampling/collection and at the lowest available temperature, at least at −20 °C but preferably at −80 °C.

3. Make sure all samples, stock, and working solutions are correctly and fully labeled. Thaw stock/working solutions and real samples shortly before use and sample preparation, respectively.

3.3 Sample Preparation

1. Vigorously vortex every sample, after thawing at room temperature, prior to sample preparation.

2. Transfer an appropriate volume (e.g., 50–200 µL) of each sample into 1.5 mL Eppendorf tubes.

3. Depending on the matrix (urine, plasma, serum), a different sample preparation procedure is recommended (*see* **Notes 5 and 6**) as explained below in Subheadings 3.3.1–3.3.3.

3.3.1 For Urine Samples

1. Dilute the sample in ratio 1:3 v/v with Millipore quality water.

2. Vortex for 5 min and centrifuge at $12,000 \times g$ or higher speeds for 5–10 min at 4 °C.

3. Transfer the clear supernatant into an LC-MS vial (n.b. 96 well plates or similar can be substituted for LC-MS vials).

4. Place the vial (well plate) into a pre-cooled (4 °C) autosampler for analysis. If analysis is not being undertaken immediately, store the vial (well plate) in a fridge (0–4 °C) if analysis will start shortly or a freezer (−20/80 °C) if analysis will be delayed for longer than a few hours.

3.3.2 For Plasma or Serum Samples

1. Add three times the sample volume of ice cold methanol or acetonitrile for protein precipitation (*see* **Note 7**).

2. Vortex for 5 min and centrifuge at $12,000 \times g$ or higher speeds for 5–10 min.

3. Transfer the clear supernatant into Eppendorf tubes.

4. Evaporate to dryness and reconstitute with water/acetonitrile 95:5 v/v to the initial volume.

5. Repeat **step 2**.

6. Transfer the clear supernatant into an LC-MS vial (well plate).

7. Place the vial (well plate) into a pre-cooled (4 °C) autosampler for analysis. If analysis is not taking place immediately, follow

the procedure described for urine in Subheading 3.3.1, **step 4**. When taking plasma/serum samples out of the freezer, it is advisable to centrifuge samples again before analysis.

3.3.3 For Quality Control Samples (QC)

1. Create a pool sample by mixing equal volumes (e.g. 20 μl) of each sample analyzed. Make an appropriate number of aliquots (a minimum of one QC for each ten samples to be analyzed and possibly up to one for every five; see below), and store them at a freezer.

2. Handle each QC as a real sample and follow the sample preparation procedure developed for the type of matrix to be analyzed.

3. If the number of the samples being analyzed is large (e.g., >1000) and the creation of a pooled sample from all the samples is impractical, consider making QC sample from only the samples contained within each analytical run being analyzed to provide a within-batch QC and have a separate "bulk matrix" QC sample also analyzed within every batch to enable between batch comparisons and normalization to be conducted [17–19].

3.3.4 For Test Mix Sample

1. Spike a QC sample with an appropriate volume of the methanolic mixture mentioned in Subheading 2.1 above, in order to achieve a final concentration of 5–10 μg/mL. It is advisable to perform this procedure by evaporating the methanolic stock solution and then reconstituting by dissolution in the QC matrix.

3.4 Analytical Sequence Preparation

1. Make sure that the order of the samples is randomized, to avoid introducing bias due to changes with individual analytical runs and between run "batch" effects. Randomization of large sample sets can be performed using the specific commands in spreadsheet programs such as Excel. QC samples should not be randomized but inserted regularly in the run sequence (e.g., one QC sample every five to ten real samples).

2. The number of QC samples placed in a sequence depends on the total number of samples analyzed and on the duration of the analysis. For a small number of samples (<100), one QC every five sample injections is recommended, and for a larger number QC, samples should represent at least 10% of the total analyzed.

3. After column equilibration, in the middle of the sequence and at the end, insert the spiked test mix solution. Standard solution injections should be avoided within the batch of test samples, as the system can be disequilibrated by the injection of non-matrix

solutions (*see* **Notes 3** and **8**). Similarly blank sample injections should be avoided.

4. A good way to improve quantification aspects in such analysis is to implement a sequence of injections of serial dilutions of the QC sample. This enables observations on saturating peaks, which would be better detected in dilute samples (e.g., 1:10 v/v). At the same time, this series of analyses facilitates a study of the response of non-saturating peaks in response to dilution.

In untargeted metabolomic studies, data analysis strategies require some steps involving raw data acquisition, normalization, scaling, and feature and peak detection in order to finally reach biomarker detection and identification.

3.5 Data Analysis

3.5.1 Data Processing

From the raw data acquired, observe the peak width, retention time, peak intensity/signal, mass accuracy, and noise intensity from indicative chromatographic peaks located at the beginning, middle, and end of the chromatogram. These can help researchers to define carefully the optimum parameters for the software (XCMS, MassLynx, etc.) to be used for peak alignment, peak peaking, and integration.

Data from the QC samples are used to evaluate the quality of analysis via the following steps: Treat QC samples as a separate group and process them alone using exactly the same parameters selected for processing the whole sample set. Peaks that appear in less than 70% or 80% of the QC samples should be omitted from the data set [14]. Apply principal component analysis (PCA), using SIMCA-P or other statistical software, on the processed data from the QC samples in order to observe any trends, such as time-related drift during the run, on these specific samples.

More details on data pre-processing can be found in Chapters 3 and 4 of this book.

3.5.2 Data Quality Evaluation

For data quality evaluation, there are some steps that should be taken into consideration. It is advisable to subject the data to these tests against preset criteria of quality as explained below.

Firstly, assess the performance of the analytical run by inspecting an overlay of full-scan chromatograms of all the replicates of the test mix samples, placed at the beginning, middle, and end of the sequence. Calculate chromatographic data, such as the retention time, peak area, and height, and determine the precision (RSD %) of each value. The acceptable limits should be less than 2% for retention time variation and less than 20–30% for peak area variation. In case that these conditions are not met, reexamine the data carefully to find any trend that might indicate system underperformance or malfunction.

The next step is to assess the data acquired from QC samples regularly placed among the samples, in order to strengthen and corroborate the initial assessment of the performance of the analytical run. Evaluate the full-scan chromatograms of all QC samples. It is expected that all QC samples show good consistency, reproducibility, and small variability, with the possible exception of the first five to ten injections (conditioning QCs). An overlay of full-scan chromatograms can provide only an estimation of analytical repeatability and system stability. A thorough check should be done on extracted ion chromatograms of various peaks distributed along the whole length of the analysis. Examined peaks should be spread in both retention time and detected mass. The precision of the retention time, the mass detected, and the signal intensity of different peaks located across the chromatogram should be performed with preset criteria (e.g., RSD% should be less than 2% for retention times, less than 20% for the area of abundant peaks and less than 30% for low intensity peaks) as mentioned above. Peaks that fail such criteria should be excluded from the data set (including data from the test samples). Evidence of poor repeatability for the QC samples should be taken as a possible reason for not accepting a run.

If there are any problems concerning the reproducibility of the data, try to estimate where the inconsistency is appearing. If it is located in a specific time window, investigate if it is worth excluding this window in the peak picking and alignment process. If the data is poor and fails the acceptance criteria, investigate the reason for the failure and repeat the analysis.

Providing that the run passes the predetermined acceptance criteria, perform principal component analysis (PCA) on all samples and QC replicates, in order to observe variations in the acquired data. PCA will immediately show similarities or differences among real and QC samples. The latter should ideally form a tight cluster (assuming that there are differences between the samples). For untargeted data, investigate the effect of different scaling modes. Applying Pareto scaling and different visualization tests helps in identifying potential underlying issues. The following steps are suggested:

1. Color QC and real samples according to analytical run order and try to identify time trends in the score plots.

2. Check for any QC outliers or for batch effect due to run order

3. Examine the QC data for any unexpected trends. If such trends are found, study additional visualization plots such as time series plots, and look for samples that are out of 2SD or 3SD thresholds. Test samples appearing as such "outliers" should not be

discarded outright but rather be thoroughly scrutinized. QC samples out of 2SD indicate that the data from the neighboring test samples should be thoroughly scrutinized and probably discarded with the samples themselves being reanalyzed.

Thereinafter, in the exported peak table data, look for any similarities in the pairs (retention time and mass) among the peaks reported, in order to assess the success of peak alignment. Then, using appropriate statistical software, create graphs with the data exported from the QC samples to evaluate the analytical performance and precision.

Figure 1 presents a schematic of the proposed validation scheme. It starts with the QC preparation, by mixing equal volumes of all the real samples (urine, plasma/serum, etc.) followed by determing the run order sequence and then injecting the QC sample periodically, e.g., every five to ten real samples during analysis. From the resulting data set, an extracted ion chromatogram (XIC) of all QC samples can be created, via data processing or data mining, where data is collected only for specific m/z values of the compounds of interest. Those results can also be used for the formation of an *analysis chart*, showing total sequence order (both QC and real samples). Data processing is complemented by statistical analysis. Repeatability and run order trends can be assessed by making QC *quality control charts* and *data tables*, evaluating %RSD values of QC's ion intensities. Features presented in most samples (~70%) with %RSD values <30% indicate a good data set. Finally, PCA score plots can be created from the examined results, providing information on the accuracy and precision of the analysis, as well as for any time trends related to the order of analysis of both QC and test samples.

4 Notes

1. The storage period for each standard in the freezer is dependent on the nature of the analyte and the solvent used for dissolution. Certain standards dissolved in water are stable for up to 20 days, or longer, at −20 °C.

2. Make sure mobile phases are filtered and sonicated to degas them properly.

3. Avoid making injections with standard solutions during the sequence so the system isn't deconditioned.

4. Always monitor and record the exact column pressure (bar or psi) at the beginning and end of a run. Rising pressure can be an indication that a column is becoming blocked and so provides an indication of the need for either replacement or cleaning.

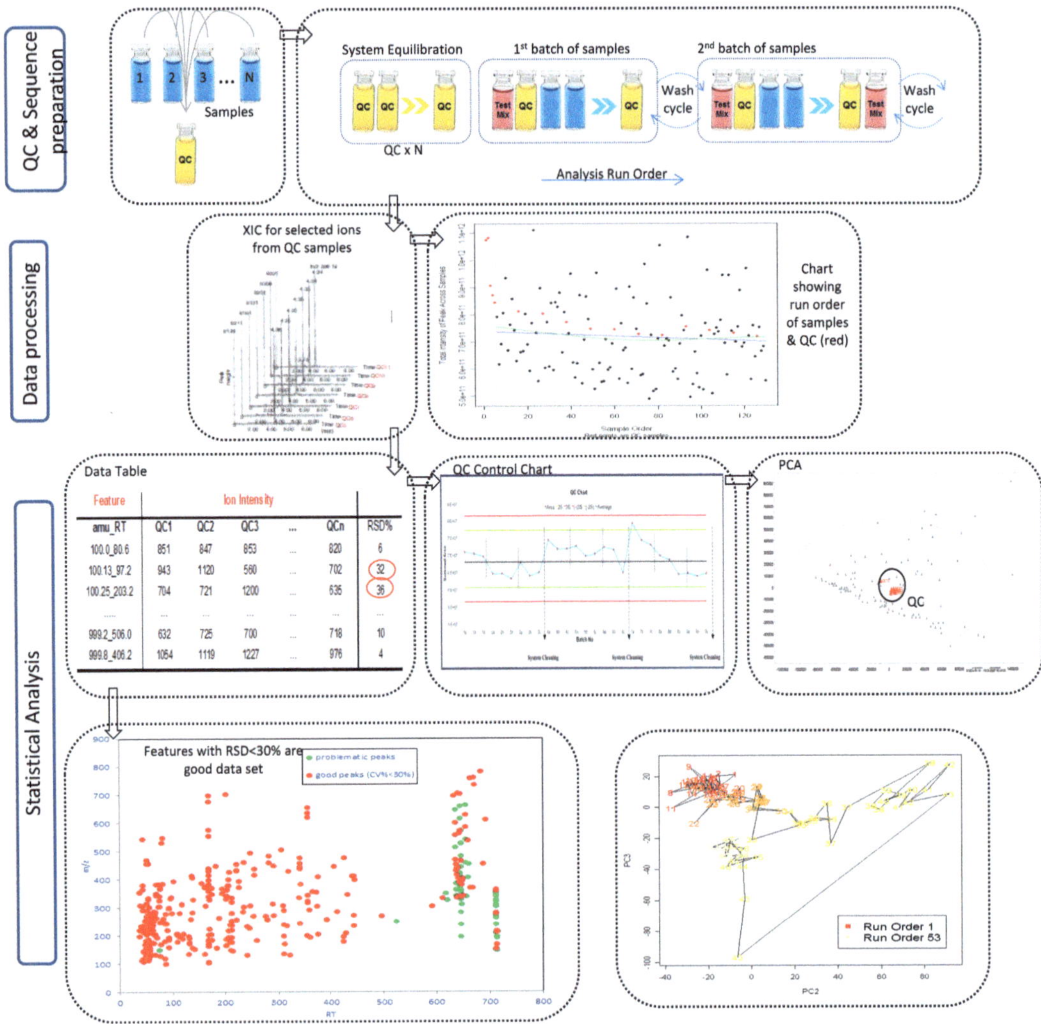

Fig. 1 Schematic of a validation scheme from QC and sequence preparation (top), to data processing (middle) to statistical analysis (bottom). From top left: QC samples are prepared by initially mixing equal volume of all the real samples. Data processing includes scrutiny of data as extracted ion chromatogram (XIC) and data processing or data mining. Trends in data should be checked across the run order. Repeatability and run order trends can be assessed by making *QC quality control charts* and *data tables*, evaluating %RSD values of QC's ion intensities. PCA score plots should exhibit tight cloud of QC samples (red in the presented plot) within the whole sample set (black dots). In the bottom of the figure, a control chart (left) shows features showing poor repeatability in red (CV >30% in QC samples) and repeatable features in green (CV <30% in QC samples). These charts are providing information for the precision of the analysis, as well as for any time trends related to the order analysis of both QC and real samples (bottom right)

5. Apply proper cleanup steps during sample preparation. Filtering samples improves the medium-/long-term integrity of the separation system and as a result may safeguard high-quality results if necessary.

6. During sample preparation, all necessary safety precautions must be taken. Make sure gloves and goggles are used and that samples are prepared under fume hoods.

7. It has been noticed that using ice cold solvents (methanol, acetonitrile) at three times the sample volume helps to provide better protein precipitation than solvents at ambient temperature.

8. In the case of large numbers of samples being analyzed, divide them into randomized batches. Carefully monitor the system performance during each batch and have appropriate cleaning cycles in between (column, cone, source, etc.). Before starting a new sequence, make sure that the system is re-equilibrated.

References

1. Nicholson JK, Lindon JC, Holmes E (1999) "Metabonomics": understanding the metabolic responses of living systems to pathophysiological stimuli via multivariate statistical analysis of biological NMR spectroscopic data. Xenobiotica 29:1181–1189. https://doi.org/10.1080/004982599238047

2. Dunn WB, Wilson ID, Nicholls AW, Broadhurst D (2012) The importance of experimental design and QC samples in large-scale and MS-driven untargeted metabolomic studies of humans. Bioanalysis 4:2249–2264. https://doi.org/10.4155/bio.12.204

3. Griffiths WJ, Koal T, Wang Y et al (2010) Targeted metabolomics for biomarker discovery. Angew Chem Int Ed 49:5426–5445. https://doi.org/10.1002/anie.200905579

4. Wilson I (2016) Methods and techniques for metabolic phenotyping. Bioanalysis 9:1–3. https://doi.org/10.4155/bio-2016-4985

5. Gika HG, Wilson ID, Theodoridis GA (2014) LC–MS-based holistic metabolic profiling. Problems, limitations, advantages, and future perspectives. J Chromatogr B 966:1–6. https://doi.org/10.1016/j.jchromb.2014.01.054

6. Gika HG, Theodoridis GA, Earll M et al (2010) Does the mass spectrometer define the marker? A comparison of global metabolite profiling data generated simultaneously via UPLC-MS on two different mass spectrometers. Anal Chem 82:8226–8234. https://doi.org/10.1021/ac1016612

7. Rainville PD, Theodoridis G, Plumb RS, Wilson ID (2014) Advances in liquid chromatography coupled to mass spectrometry for metabolic phenotyping. TrAC Trends Anal Chem 61:181–191. https://doi.org/10.1016/j.trac.2014.06.005

8. Gika HG, Theodoridis GA, Earll M, Wilson ID (2012) A QC approach to the determination of day-to-day reproducibility and robustness of LC-MS methods for global metabolite profiling in metabonomics/metabolomics. Bioanalysis 4:2239–2247. https://doi.org/10.4155/bio.12.212

9. Begou O, Gika HG, Wilson ID et al (2017) Hyphenated MS-Based Targeted approaches in metabolomics. Analyst 142(17):3079–3100

10. Theodoridis G, Gika HG, Wilson ID (2011) Mass spectrometry-based holistic analytical approaches for metabolite profiling in systems biology studies. Mass Spectrom Rev 30:884–906. https://doi.org/10.1002/mas.20306

11. Gika HG, Zisi C, Theodoridis G et al (2016) Protocol for quality control in metabolic profiling of biological fluids by U(H)PLC-MS. J Chromatogr B Analyt Technol Biomed Life Sci 1008:15–25. https://doi.org/10.1016/j.jchromb.2015.10.045

12. Gorrochategui E, Jaumot J, Lacorte S et al (2016) Data analysis strategies for targeted and untargeted LC-MS metabolomic studies: overview and workflow. TrAC Trends Anal Chem 82:425–442. https://doi.org/10.1016/j.trac.2016.07.004

13. Naz S, Vallejo M, García A et al (2014) Method validation strategies involved in non-targeted metabolomics. J Chromatogr A 1353:99–105. https://doi.org/10.1016/j.chroma.2014.04.071

14. Bijlsma S, Bobeldijk I, Verheij ER et al (2006) Large-scale human metabolomics studies: a strategy for data (pre-) processing and validation. Anal Chem 78:567–574. https://doi.org/10.1021/ac051495j

15. Want EJ, Wilson ID, Gika H et al (2010) Global metabolic profiling procedures for urine using UPLC-MS. Nat Protoc 5:1005–1018. https://doi.org/10.1038/nprot.2010.50

16. Gika HG, Theodoridis GA, Wingate JE et al (2007) Within-day reproducibility of an HPLC-MS-based method for metabonomic analysis: application to human urine. J Proteome Res 6:3291–3303. https://doi.org/10.1021/pr070183p

17. Zelena E, Dunn WB, Broadhurst D et al (2009) Development of a robust and repeatable UPLC–MS method for the long-term metabolomic study of human serum. Anal Chem 81:1357–1364. https://doi.org/10.1021/ac8019366

18. Dunn WB, Broadhurst D, Begley P, Human Serum Metabolome (HUSERMET) Consortium (2011) Procedures for large-scale metabolic profiling of serum and plasma using gas chromatography and liquid chromatography coupled to mass spectrometry. Nat Protoc 6:1060–1083. https://doi.org/10.1038/nprot.2011.335

19. Lewis MR, Pearce JTM, Spagou K et al (2016) Development and application of ultra-performance liquid chromatography-TOF MS for precision large scale urinary metabolic phenotyping. Anal Chem 88:9004–9013. https://doi.org/10.1021/acs.analchem.6b01481

20. Veselkov KA, Vingara LK, Masson P et al (2011) Optimized preprocessing of ultra-performance liquid chromatography/mass spectrometry urinary metabolic profiles for improved information recovery. Anal Chem 83:5864–5872. https://doi.org/10.1021/ac201065j

21. Michopoulos F, Theodoridis G, Smith CJ, Wilson ID (2010) Metabolite profiles from dried biofluid spots for metabonomic studies using UPLC combined with oaToF-MS. J Proteome Res 9:3328–3334. https://doi.org/10.1021/pr100124b

22. Ramautar R, Nevedomskaya E, Mayboroda OA et al (2011) Metabolic profiling of human urine by CE-MS using a positively charged capillary coating and comparison with UPLC-MS. Mol BioSyst 7:194–199. https://doi.org/10.1039/c0mb00032a

Chapter 3

Data Treatment for LC-MS Untargeted Analysis

Samantha Riccadonna and Pietro Franceschi

Abstract

Liquid chromatography-mass spectrometry (LC-MS) untargeted experiments require complex chemometrics strategies to extract information from the experimental data. Here we discuss "data preprocessing", the set of procedures performed on the raw data to produce a data matrix which will be the starting point for the subsequent statistical analysis. Data preprocessing is a crucial step on the path to knowledge extraction, which should be carefully controlled and optimized in order to maximize the output of any untargeted metabolomics investigation.

Key words Preprocessing, Peak picking, Retention time correction, Metadata, Quality check, Missing values

1 Introduction

Liquid chromatography-mass spectrometry (LC-MS) is an established and widely used analytical technique and combines the separation potential of LC with the ability of MS to quantify the ion intensity [1]. From a data scientist point of view, LC-MS produces "three-dimensional" datasets where each ion generated in the ionization source is characterized by m/z value, retention time (rt) and intensity.

While in targeted applications selected ions are used to quantify the analytes of interest, untargeted investigations deal with the full set of ionic signals trying to characterize all the detectable analytes including chemical unknowns [2]. This potential, however, comes at a price because the knowledge extraction process is not straightforward and heavily relies on bioinformatics [3].

A schematic representation of the data analysis workflow maximizing the process of knowledge mining in an untargeted LC-MS investigation is presented in Fig. 1: the path involves a constant feedback between experimental design (DoE), data processing and statistical analysis. An extensive treatment of all the steps goes beyond the scope of this chapter and here we will focus on the

Georgios A. Theodoridis et al. (eds.), *Metabolic Profiling: Methods and Protocols*, Methods in Molecular Biology, vol. 1738, https://doi.org/10.1007/978-1-4939-7643-0_3, © Springer Science+Business Media, LLC, part of Springer Nature 2018

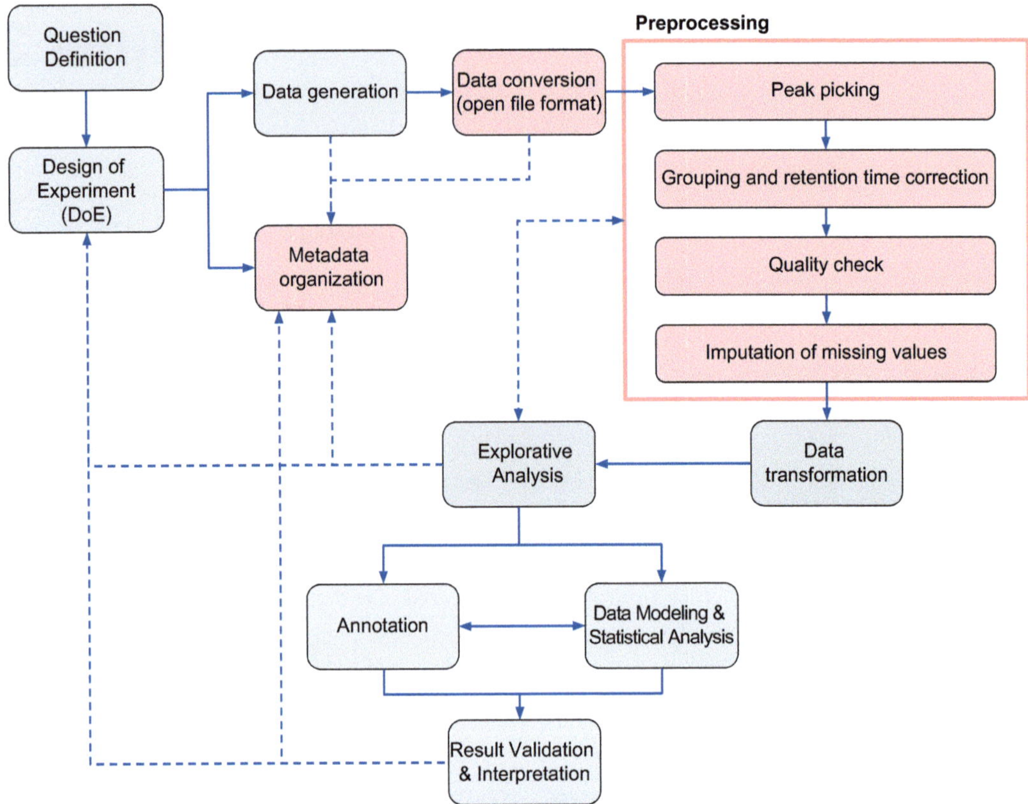

Fig. 1 Data analysis workflow of a typical LC-MS untargeted experiment: the path from data to knowledge is usually represented as linear, but it is more likely to have many feedback loops (dashed lines). Only the red boxes are detailed in the text

so-called data preprocessing. The objective of this phase is to summarize the raw analytical data into a two-dimensional matrix with samples on the rows and ions—usually indicated as *features*—on the columns. Our discussion will also touch data organization, because this step plays a fundamental role in guaranteeing the interoperability and reproducibility of the whole data analysis process [4]. Considering the high pace of evolution of bioinformatics, a pure "how-to" guide would be outdated quite fast. Thus we try to include also some general considerations valid in a wider variety of contexts. In parallel, to give practical guidance to the researcher, we illustrate how the workflow can be implemented in the statistical environment R [5], which is obviously not the only possible solution [6–11]. We believe that a general understanding of the key aspects of the problem can be of value beyond the specific solution chosen to perform data analysis.

2 Software

1. **Metadata organization**. Spreadsheets to inspect/fill tables. The ISA metadata model to encode the information [4, 12]. The Risa Bioconductor package [13] and the ISAcreator application [14] to fill in and read ISA-Tab files.

2. **Data conversion**. Vendor software allowing data export in open file format (*see* **Note 1**) or the cross-platform ProteoWizard toolkit [15, 16].

3. **Processing (peak picking, grouping and retention time correction, quality check, imputation of missing peaks)**. The xcms [17–19] Bioconductor package, which is a popular solution for processing raw chromatography-MS data.

4. **Example dataset**. Spiked-apple dataset available from MetaboLights (MTBLS59) [20], downloaded and unzipped in a directory named "MTBLS59" within the R working directory [20]. Briefly, the dataset contains the results of the analysis of 40 apple extracts divided in four classes that were spiked with a set of known compounds. The extracts were analysed by UPLC-QTOF—MS in positive and negative mode.

3 Methods

3.1 Metadata Organization

The term "experiment metadata" refers to the set of information describing the samples and the analytical pipeline (*see* **Note 2**). Organizing and storing a detailed record of the sample metadata (*see* **Note 3**) is crucial to perform a reliable interpretation of the results, evaluating the presence (and the impact) of possible confounding factors.

1. Collect and organize (possibly with the help of a spreadsheet) all the important properties of the samples (e.g. treatment type, day of collection, origin, etc.). Keep track also of the sample specific analytical details.

2. Run ISAcreator to generate your ISA-Tab files (*see* **Note 4**). The tool will support the description, through commonly accepted "ontologies", of the general aim of your experiment ("Study"), the analytical platform ("Study Assay"), the DoE ("Study Design Descriptors", "Study Factors"), the analytical information ("Study Protocols") and the details about each sample ("Sample Definitions"). Handling large datasets with ISAcreator may not be easy, in particular if one wants to automatically fill some of the required fields. In this case, an alternative choice is to use ISAcreator only to generate the backbone of the ISA-Tab files and to fill in the sample details either by a spreadsheet or a scripting language (e.g. R, python, etc.). In R,

ISA-Tab files can be managed directly by using the Risa package, which is based on the definition of the ISAtab-class. If the ISA-Tab files are stored into a folder named *"ISA_info"* in the R working directory, they can be read into R by typing

```
mydata<-readISAtab(path="ISA_info")
```

The sample information can then be modified using the function updateAssayMetadata, and, once finished, the ISA-Tab files can be saved into a new folder *"ISA_exp_description"* with the command (*see* **Note 5**):

```
write.ISAtab(mydata,path="ISA_exp_description")
```

3. Perform the final consistency checks in ISAcreator.

4. Create the final archive containing the raw data and their description with ISAcreator.

3.2 Data Conversion (See Note 6)

In general, the machine software can be used to perform this task (*see* **Note 7**). At present, the open source alternative is the ProteoWizard "msconvert" (*see* **Note 8**), which is available as a command line tool or as a point-and-click application for Windows (MSconvertGUI) (*see* **Note 9**).

1. Open a terminal or a command window and run msconvert into the raw data directory. As an example, to convert vendor-specific raw data (here .RAW data files) into mzML files and store them into the *"open_data"* directory use:

```
msconvert *.RAW -o open_data.
```

Alternatively, the use of the GUI application is self-explanatory. Also in this case, batch mode conversion of a set of files is allowed.

3.3 Processing

Each metabolite produces at least one peak in the (m/z, rt) plane. To detect metabolites, an automatic algorithm has to be able to identify such peaks and to distinguish them from electrical or chemical noise [7, 21]. This process of peak identification (peak picking) has to be performed automatically on all the samples under analysis. Subsequently, the peak lists identified in each sample have to be joined together in a final list of features. To do that, it is necessary to account for m/z and rt shifts from sample to sample [22]. The measured m/z values can be different because even the most advanced mass spectrometer is not absolutely precise. Rt shifts, instead, are unavoidable characteristics of chromatography. Technically speaking, a "retention time correction" step is implemented to compensate for the shifts in rt, and, afterwards, the features found in the different samples are "grouped" (across samples) in a final consensus list. All the software solutions (commercial or open source) implement different strategies to perform these steps [23, 24].

3.4 Preprocessing: Peak Picking

The characteristics that distinguish a peak from the noise are intensity, width and shape. Different algorithms handle them in different ways [7, 21]. Most importantly, not only the algorithm but also the specific parameters of each algorithm greatly affect the results (*see* **Note 10**).

1. Choose the peak-picking algorithm on the basis of the characteristics of the analytical pipeline [25, 26]. In xcms several choices are already available (*see* **Note 11**). The type of instrumental setting used to analyse MTBLS59 suggests using the centWave algorithm [18].

2. Identify the critical parameters of the algorithm and possibly link them with the characteristics of the analytical pipeline. For centWave, for example, the most important parameters to be considered are ppm, which is determined by the mass accuracy and is used to determine the "reliability" of each mass trace; prefilter, which has to be set considering the minimum intensity (I) of a true signal (at the single spectrum level) and also in how many consecutive scans (k) that signal should be detected; peakwidth, which deals with the range of the acceptable chromatographic peak-widths (in seconds); and snthresh, which defines a cut-off for distinguishing signal from noise (signal-to-noise ratio).

3. Check on a representative sample the results of the peak picking (*see* **Note 12**). For instance, to run centWave with default parameters on a specific file, issue

```
test_sample <- xcmsSet(files="MTBLS59/apple_
control_neg_001.CDF", method="centWave")
```

   To visually inspect the outcomes of this step, plotting the peaks in the *m/z*, rt plane, issue

```
mypeaks<-peaks(test_sample)
plot(mypeaks[,c("rt","mz")])
```

4. Optimize the peak-picking parameters until the result is satisfactory (*see* **Note 13**).

5. Perform the peak picking on the whole dataset with the selected set of parameters. The final xcmsSet object (*see* **Note 11**) will contain the full list of the peaks detected in all the samples, with some additional information on the mass and retention time ranges. The complete analysis (with the parameters set as suggested in [26]) can be run issuing:

```
cdffiles <- list.files(path="MTBLS59",pattern=
   "CDF",

full.names = TRUE)

myfiles<-grep("neg",cdffiles, value=TRUE)[1:40]

xs<-xcmsSet(files=myfiles,method="centWave",
   ppm = 15, prefilter = c(0,0), peakwidth =
   c(5,20), snthresh=1)
```

3.5 Preprocessing: Retention Time Correction and Grouping (See Note 14)

1. Choose the retention time correction algorithm among the available options (*see* **Note 11**). In this example we use obiwarp [26], which is based on dynamic time warping, an algorithm used to find the best stretching of the time dimension of two time series to make them as similar as possible [27]. LC-MS data are multidimensional, so in obiwarp the similarity is determined taking into account the full spectral information.

2. Determine the more relevant parameters of the algorithm. In obiwarp the relevant parameters are the choice of the reference sample for the rt warping (center); the m/z bins used for retrieving the spectra (profStep); the spectral similarity function (distFunc); and the penalizations on the warping optimization (gapInit and gapExtend). With the default parameters (which work reasonably well in most of the cases), the reference sample is the one containing the largest number of peaks. The retention time correction can be performed with the following command.

```
xsr <- retcor(xs, method="obiwarp")
```

3. Choose the "grouping" algorithm or, in other terms, the procedure which matches peaks with similar *m/z* and rt found in different samples. Among the possible choices, we rely on the default density-based solution [17].

4. Choose the grouping parameters according to the analytical pipeline: bw defines the retention time window used for the density estimation and mzwid accounts for small differences in the *m/z* value of the corresponding peaks detected in the different samples. To avoid the inclusion in the final list of features peaks detected only in a small number of samples (which could be, e.g. artefacts), it is possible to keep only groups which contain peaks detected in at least minsamp samples (in at least one sample class). Again, we set the parameters as suggested in [26] and minsamp equal to 2:

```
xsr <- group(xsr, bw = 2, mzwid = 0.025,
    minsamp = 2)
```

3.6 Preprocessing: Quality Check

At the end of the preprocessing, it is extremely important to check the outcomes of the overall process. This can be done implementing a series of QC plots (see Chapter 2). Firstly, a global evaluation of the retention time correction procedure helps in detecting unwanted extreme deformations of the original time scale. Secondly, compounds known to be present in the samples should be correctly detected and aligned on the whole dataset.

1. Visually inspect the retention time correction. In xcms this can be done using the dedicated function.

```
plotrt(xsr)
```

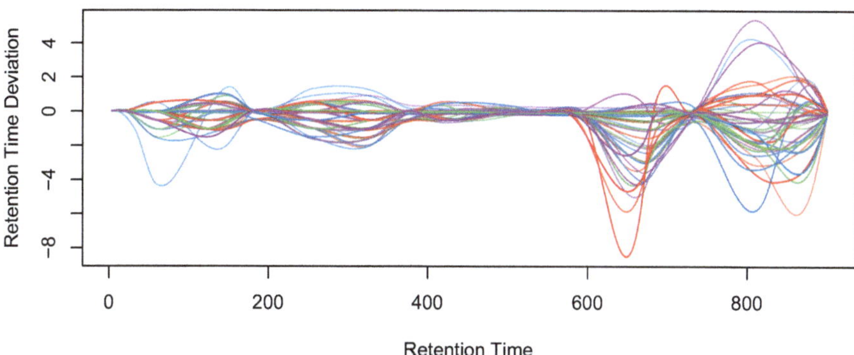

Fig. 2 Outcome of the retention time correction performed on the MTBLS59 data as obtained by plotrt, after minor visual parameters customization. Each line represents one sample and is coloured by sample class. The retention time correction is different for each sample, and it is not linear in time

The situation for the MTBLS59 data is presented in Fig. 2 which shows the retention time deviation (in seconds) for each sample along the chromatographic retention time. Usually, the first and the last part of the chromatography require the heaviest corrections. However, in the example, the greatest correction has been applied to one of the red samples and is of about 8 s. Considering that here we are dealing with UPLC, the correction is acceptable indicating a satisfactory reproducibility in the chromatography. As a rule of thumb, the maximum correction should be comparable to the chromatographic peak width in that particular rt range.

2. Inspect the extracted ion traces (EICs) associated with compounds known to be present in the samples. In presence of a good preprocessing, they should look nicely aligned across the samples. In addition, a corresponding feature (group) has to be present in the final list of features. In the case of the apple extract in MTBLS59, possible metabolites to check are quercetin-3-glucoside and quercetin-3-galactoside, both showing a strong ionic signal at m/z 463.087 (in negative ion mode) [20] at around 415 s. The EICs in the vicinity of their peak can be extracted by issuing:

```
mzrange<-cbind("mzmin"=463.087-0.005,
"mzmax"=463.087+0.005)
rtrange<-cbind("rtmin"=405,"rtmax"=430)
myeic_corrected<-getEIC(xsr,mzrange=mzrange,
rtrange=rtrange,rt="corrected")
plot(myeic_corrected)
```

The position of the corresponding feature in the groups can be extracted with the following commands:

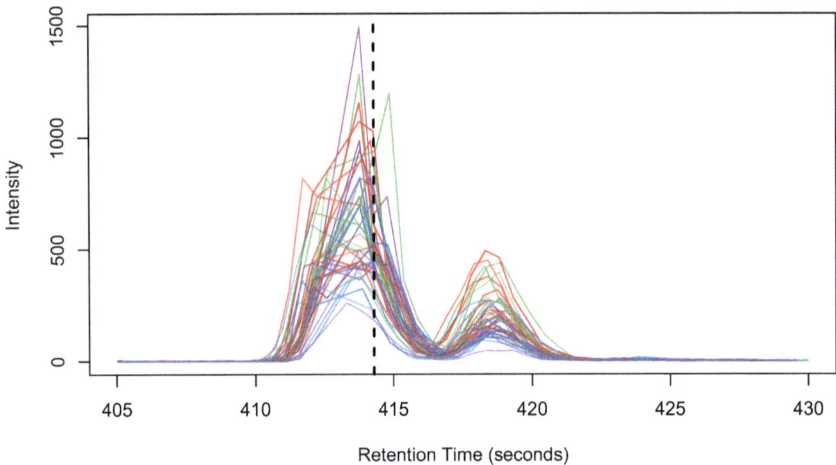

Fig. 3 Quality check on the processing procedure using the quercetin-3-glucoside and quercetin-3-galactoside in negative ionization, which are expected to be present. Minor visual parameters customization has been performed compared to the base command reported in the text. Samples are coloured by sample class. The black dashed line corresponds to the identified the peak group

```
mygroups<-groups(xsr)
mycompound<-mygroups[mygroups[,"rtmed"]>405
& mygroups[,"rtmed"]<430
& mygroups[,"mzmed"]>463.082
& mygroups[,"mzmed"]<463.092,"rtmed"]
abline(v=mycompound)
```

The resulting plot is shown in Fig. 3. It is clear that the EICs are correctly aligned (*see* **Note 15**). In this case, however, xcms finds only one feature associated with both compounds because of the initial choices of the peak-picking algorithm (*see* **Note 16**).

3.7 Preprocessing: Imputation of Missing Peaks

It could happen that peaks belonging to a subset of samples are missing from "their" groups. This is correct if the corresponding analyte is not present in a specific sample, but it could also be the result of an error in the peak-picking phase. These missing peaks produce missing values in the data matrix, which are usually tricky to handle during statistical analysis. Several options exist to "impute" missing values (which are common also in other -*omic* technologies). The clever solution available in xcms is implemented by the function fillPeaks, which goes back to the raw data integrating the signal measured in the area corresponding to the "missing" peak. If the peak was wrongly missed, fillPeaks will recover it, otherwise that missing value will be filled with the sum of the noise.

1. Run the imputing function issuing:

```
xsf <- fillPeaks(xsr)
```

3.8 Data Matrix Extraction

Extract the final two-dimensional data matrix with samples on the rows and features on the columns, using as intensity the integral of the area under the peak:

```
mat<-t(groupval(xsf,value="into"))
```

The m/z and rt details about the features are encoded in the mygroups object generated at Subheading 3.6, **step** 2.

4 Notes

1. Raw data are usually stored in vendor proprietary formats that can normally be read only by proprietary software. A close data format hampers data comparison and sharing, which is instead made possible adopting "open" data formats with publicly available specifications. Examples of open formats available for LC-MS are mzML [28], which is supported by the proteomics standards initiative, mz5 [29], mzDB [30] and netCDF.

2. Sample metadata often mirror the factors included in the experimental design (DoE) (type of treatment, gender, etc.), but they also include key characteristics of the samples. They have to be considered during the design of the analytical work-flow, in order to ensure proper randomization of the samples [31–37].

3. The use of a standardized language and/or ontologies is essential to profit as much as possible from the experiment, allowing result comparison, integration and sharing [4, 12]. The compliance with ISA standards is also a requirement for submitting experimental data (both raw data and metadata) to MetaboLights, an open-access and general purpose repository for metabolomics data [38].

4. An ISA-Tab investigation comprises a set of tab-separated files which record all the relevant metadata about the samples and the analytical pipeline. A step-by-step tutorial on how to use ISAcreator can be found on the MetaboLights website (https://www.ebi.ac.uk/metabolights/help).

5. For additional arguments and details on the functions, please refer to the package documentation or to its vignette. (https://www.bioconductor.org/packages/release/bioc/html/Risa.html)

6. It is worth mentioning that the data conversion to an open data format is an optional step. In particular, it is not manda-tory when the analyses are performed with the vendor soft-

ware. On the contrary, it is mandatory when an open source solution is preferred. It is important to know that not all the relevant analytical metadata are usually included in the open source version of the data (for instance, the analytical column temperature). To avoid information loss, it is then necessary to store a copy of the original "closed" files which can be inspected only with the vendor software. This fact has to be taken into consideration in the planning of the data storage infrastructure.

7. Examples of data conversion can be found in [9].

8. The list of open and closed formats handled by ProteoWizard is available on the project website (http://proteowizard. sourceforge.net/tools.shtml).

9. To perform the conversion, the vendor-specific Windows dynamic link libraries (DLLs) are needed for reading the raw data files and the sample metadata. Such DLLs are distributed either as part of the specific vendor software or within the Windows version of the ProteoWizard suite.

10. Automatic peak detection usually leads to suboptimal solutions when compared to visual inspection (e.g. it could happen that a peak is instead just noise). Nevertheless, visual inspection will introduce subjectivity and not reproducible results (besides the fact that it is a very time-consuming task). The advantage of automatic systems is that they always apply the same "approach" impartially. Thus given the "approach" (or, more precisely, the algorithm and its parameters), the outcome of the procedure is determined. The algorithms used to perform all the preprocessing tasks should be flexible, but an excess of flexibility can produce unwanted artefacts (wrongly matched features). Therefore to get the maximum from the data, it is necessary to match this flexibility with the real analytical characteristics of the experiment (which should be also optimized!). Given the importance of the parameter tuning phase, automatic optimization tools has also been developed [39].

11. The characteristics of the different algorithms are discussed in the xcms vignette and manual, both available from the Bioconductor website (https://bioconductor.org/packages/ release/bioc/html/xcms.html).

12. Since the chosen values for the peak-picking parameters will be applied to the whole dataset, it is important that the reference is representative of the sample complexity. The more representative the sample is, the better the final outcome of the analysis will be. If quality control (QC) samples are a pool of samples, they are usually a good choice.

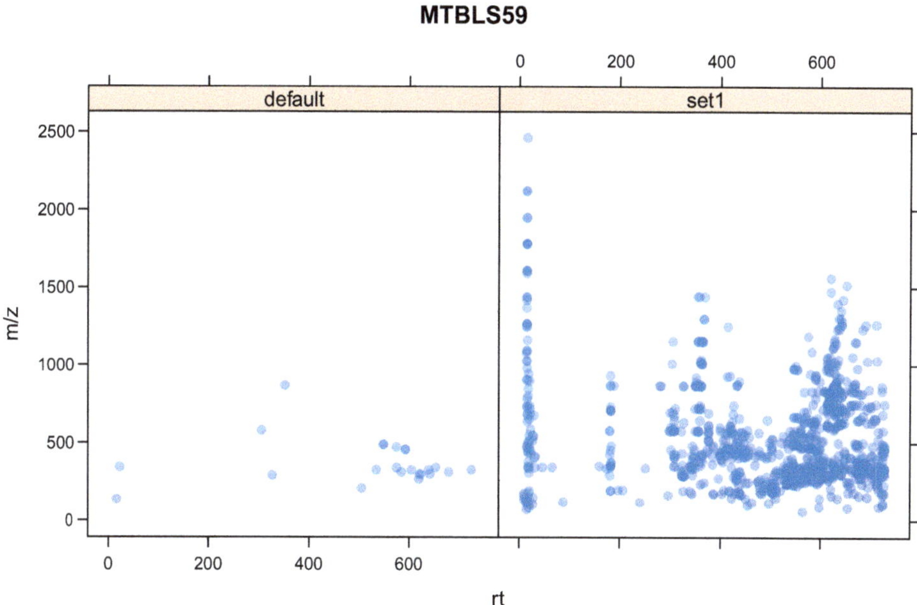

Fig. 4 Peak-picking outcome on the MTBLS59 dataset: (left) the peak detection is performed using the default settings for centWave (right) or the set of parameters defined in Methods Subheading 3.4, *step 5*

13. The signature of a good peak picking is a structured organization of points in the space defined by the retention time (rt) on the abscissa and *m/z* on the ordinate. Since each chemical produces several *m/z*'s signals at the same retention time, the points should be arranged in "vertical stripes". Dots spread uniformly in the plane or clear horizontal patterns are the symptoms of a problem in peak picking or in the chromatography itself. The number of detected peaks can also be an interesting indicator of a sensible choice of the parameters, as shown in Fig. 4. The situation on the left panel of Fig. 4 is obviously not satisfactory.

14. In some cases an additional grouping step can be necessary before retention time correction.

15. Actually a minor retention time correction has been applied here: the compound elutes within 405–430 s which correspond to a correction smaller than 2 s can be noted from Fig. 3. The EICs for the uncorrected measures can be obtained using

```
myeic_raw<-getEIC(xsr,mzrange=mzrange,rtrang
    e=rtrange,
rt="raw")
```

16. The choice of the peak-picking algorithm and of its parameters obviously limits the range of compounds that can be detected. In many cases, the samples will contain analytes eluting with diverse chromatographic peak shapes. An automatic algorithm

will never be able to tackle all the possible analytical situations. The optimal preprocessing finds the "best" and less critical compromise between flexibility and reliability. The case presented in Fig. 3 is particularly critical because the two isomers elute almost at the same time and their chromatographic peaks are not well separated. The presence of a single feature for these distinct metabolites would make impossible to rely on these settings to study a biological process which affect the balance between these two metabolites.

References

1. Ardrey RE (2003) Liquid chromatography – mass spectrometry: an introduction. John Wiley & Sons, Chichester, UK

2. Patti GJ, Yanes O, Siuzdak G (2012) Innovation: metabolomics: the apogee of the omics trilogy. Nat Rev Mol Cell Biol 13:263–269

3. Alonso A, Marsal S, Julià A (2015) Analytical methods in untargeted metabolomics: state of the art in 2015. Front Bioeng Biotechnol 3:23

4. Sansone S-A, Rocca-Serra P, Field D et al (2012) Toward interoperable bioscience data. Nat Genet 44:121–126

5. R Core Team (2016) R: a language and environment for statistical computing. R Foundation for Statistical Computing, Vienna, Austria. www.R-project.org

6. De Vos RCH, Moco S, Lommen A et al (2007) Untargeted large-scale plant metabolomics using liquid chromatography coupled to mass spectrometry. Nat Protoc 2:778–791

7. Rafiei A, Atefeh R, Lekha S (2014) Comparison of peak-picking workflows for untargeted liquid chromatography/high-resolution mass spectrometry metabolomics data analysis. Rapid Commun Mass Spectrom 29:119–127

8. Yu T, Park Y, Johnson JM, Jones DP (2009) apLCMS–adaptive processing of high-resolution LC/MS data. Bioinformatics 25:1930–1936

9. Gorrochategui E, Jaumot J, Tauler R (2015) A protocol for LC-MS metabolomic data processing using chemometric tools. Protocol Exchange. https://doi.org/10.1038/protex.2015.102

10. Katajamaa M, Oresic M (2007) Data processing for mass spectrometry-based metabolomics. J Chromatogr A 1158:318–328

11. Lange E, Tautenhahn R, Neumann S, Gröpl C (2008) Critical assessment of alignment procedures for LC-MS proteomics and metabolomics measurements. BMC Bioinformatics 9:375

12. Rocca-Serra P, Salek RM, Arita M et al (2016) Data standards can boost metabolomics research, and if there is a will, there is a way. Metabolomics 12:14

13. González-Beltrán A, Neumann S, Maguire E et al (2014) The Risa R/bioconductor package: integrative data analysis from experimental metadata and back again. BMC Bioinformatics 15(Suppl 1):S11

14. Rocca-Serra P, Brandizi M, Maguire E et al (2010) ISA software suite: supporting standards-compliant experimental annotation and enabling curation at the community level. Bioinformatics 26:2354–2356

15. Kessner D, Chambers M, Burke R et al (2008) ProteoWizard: open source software for rapid proteomics tools development. Bioinformatics 24:2534–2536

16. Chambers MC, Maclean B, Burke R et al (2012) A cross-platform toolkit for mass spectrometry and proteomics. Nat Biotechnol 30:918–920

17. Smith CA, Want EJ, O'Maille G et al (2006) XCMS: processing mass spectrometry data for metabolite profiling using nonlinear peak alignment, matching, and identification. Anal Chem 78:779–787

18. Tautenhahn R, Böttcher C, Neumann S (2008) Highly sensitive feature detection for high resolution LC/MS. BMC Bioinformatics 9:504

19. Benton HP, Want EJ, Ebbels TMD (2010) Correction of mass calibration gaps in liquid chromatography-mass spectrometry metabolomics data. Bioinformatics 26:2488–2489

20. Franceschi P, Masuero D, Vrhovsek U et al (2012) A benchmark spike-in data set for biomarker identification in metabolomics: spike-in metabolomics apple data set. J Chemom 26:16–24

21. Zhang J, Gonzalez E, Hestilow T et al (2009) Review of peak detection algorithms in liquid-chromatography-mass spectrometry. Curr Genomics 10:388–401

22. Aberg KM, Alm E, Torgrip RJO (2009) The correspondence problem for metabonomics datasets. Anal Bioanal Chem 394:151–162

23. Smith R, Ventura D, Prince JT (2015) LC-MS alignment in theory and practice: a comprehensive algorithmic review. Brief Bioinform 16:104–117

24. Koch S, Bueschl C, Doppler M et al (2016) MetMatch: a semi-automated software tool for the comparison and alignment of LC-HRMS data from different metabolomics experiments. Meta. https://doi.org/10.3390/metabo6040039

25. Brodsky L, Moussaieff A, Shahaf N et al (2010) Evaluation of peak picking quality in LC-MS metabolomics data. Anal Chem 82:9177–9187

26. Patti GJ, Tautenhahn R, Siuzdak G (2012) Meta-analysis of untargeted metabolomic data from multiple profiling experiments. Nat Protoc 7:508–516

27. Prince JT, Marcotte EM (2006) Chromatographic alignment of ESI-LC-MS proteomics data sets by ordered bijective interpolated warping. Anal Chem 78:6140–6152

28. Martens L, Chambers M, Sturm M et al (2011) mzML–a community standard for mass spectrometry data. Mol Cell Proteomics 10:R110.000133

29. Wilhelm M, Kirchner M, Steen JAJ, Steen H (2012) mz5: space- and time-efficient storage of mass spectrometry data sets. Mol Cell Proteomics 11:O111.011379

30. Bouyssié D, Dubois M, Nasso S et al (2015) mzDB: a file format using multiple indexing strategies for the efficient analysis of large LC-MS/MS and SWATH-MS data sets. Mol Cell Proteomics 14:771–781

31. Krzywinski M, Altman N (2014) Points of significance: designing comparative experiments. Nat Methods 11:597–598

32. Krzywinski M, Altman N (2014) Points of significance: analysis of variance and blocking. Nat Methods 11:699–700

33. Krzywinski M, Altman N, Blainey P (2014) Points of significance: nested designs. Nat Methods 11:977–978

34. Krzywinski M, Altman N (2014) Points of significance: two-factor designs. Nat Methods 11:1187–1188

35. Altman N, Krzywinski M (2014) Points of significance: sources of variation. Nat Methods 12:5–6

36. Blainey P, Krzywinski M, Altman N (2014) Points of significance: replication. Nat Methods 11:879–880

37. Altman N, Krzywinski M (2015) Points of significance: split plot design. Nat Methods 12:165–166

38. Haug K, Salek RM, Conesa P et al (2013) MetaboLights–an open-access general-purpose repository for metabolomics studies and associated meta-data. Nucleic Acids Res 41:D781–D786

39. Libiseller G, Dvorzak M, Kleb U et al (2015) IPO: a tool for automated optimization of XCMS parameters. BMC Bioinformatics 16:118

Chapter 4

Bio- and Chemoinformatics Approaches for Metabolomics Data Analysis

Michael Witting

Abstract

Metabolomics data analysis includes several repetitive tasks, including data sorting, calculation of exact masses or other physicochemical properties, or searching for identifiers in different databases. Several of these tasks can be automated using command line tools or short scripts in different scripting languages like Perl, Python, or R. This chapter presents simple solutions and short scripts written in R that can be used for the interaction with specific web services or for the calculation of physicochemical properties or molecular formulae.

Key words R, isotope pattern, Formula calculation, Physicochemical properties, Command line, Web service, Identifier conversion

1 Introduction

Metabolomics data analysis is very diverse in its nature and ranges from primary data pre-processing, e.g., alignment, peak picking, and normalization, to biological interpretation of obtained statistical results. Automation in metabolomics is a key point for a large-scale analysis. While the instrumental part today is largely automated, and peak pickers in current software can handle several hundreds to thousands of LC-MS runs, the final data analysis still has to be performed manually to a certain degree. However, different steps in data preparation or annotation can be performed in an automated manner using scripting languages like Perl, Python, or R. These languages often require only a minimal understanding of programming to perform basic tasks of data manipulation. R is a prominent example, extensively used in metabolomics, e.g., the processing package XCMS was developed in R [1, 2] (*see* Chapter 3). However, automated data preparation should not only focus on the initial steps of pre-processing but should also be included in later ones, especially for time-consuming, repetitive tasks. In metabolite identification, different physicochemical properties like

Georgios A. Theodoridis et al. (eds.), *Metabolic Profiling: Methods and Protocols*, Methods in Molecular Biology, vol. 1738, https://doi.org/10.1007/978-1-4939-7643-0_4, © Springer Science+Business Media, LLC, part of Springer Nature 2018

exact mass or the water/octanol partition coefficient can be calculated for the comparison with measured masses and the stratification of potential candidate structures. For example, a molecule with a very low logP value cannot be a candidate structure for a metabolite eluting late in a reversed-phase separation. This approach has been used by several groups and allows a fast reduction of candidate lists [3, 4]. Different online tools or freeware solutions are available for the calculation of different molecular properties and are usually used for single molecules drawn or uploaded one at a time, which in case of several hundreds of metabolites is tedious and time-consuming. Furthermore, metabolite identification is facing a new frontier, de novo description of potentially hundreds to several thousands of new molecules with unknown or only partially known structure. The usual first step is the calculation of molecular formulas from exact mass following specific rules, known as the seven golden rules [5]. Several algorithms for the calculation of molecular formulas and their validation using isotope patterns have been described and are used in the field [6–8]. At a certain stage, new lead structures for novel metabolites have to be generated, which need the enumeration of all possible chemical structures from a single molecular formula. Different software tools for this tasks exist, e.g., MOLGEN and the Open Molecule Generator (OMG) and its derivative Parallel Molecule Generator (PMG) [9–11].

Lastly, reporting of metabolite structures in an automatically searchable manner is extremely important. Current metabolite databases contain several hundreds of thousand metabolite structures, each one with a unique identifier in each individual database. In order to avoid overlap between different database searches or not to report duplicate structures, unique identifiers are important, whereas the chemical structure of a metabolite is the most unique identifier possible. Therefore, it should always be reported with metabolite identifications. However, images of structures cannot be read by automatic systems. Different string representations of molecules have been invented, including the SMILES and InChI system. For larger molecules, these strings can be quite lengthy. In case of InChI, a short representation of the InChI key has been invented, which represents a hashed version of the long InChI code. Fiehn and Kind reported that reporting of chemical structures in databases, as well as in publications, is an important step to unify approaches in metabolomics [12]. However, machine readable structure representation like SMILES or InChI keys is hardly human readable. The same is true for IUPAC names of molecules if the molecule is getting larger. Therefore, often trivial names or identifiers of common biological databases, like KEGG, HMDB, Lipid Maps, or ChEBI, are used. In contrast to structures, they can be ambiguous. For example, acetate is usually used as a name for acetic acid, since it is the form in which it is present in cells. In KEGG, acetate and acetic acid are stored under the same KEGG id but have different accession numbers in ChEBI. Cross mapping of different database identifiers is therefore important and has to be updated on a regular basis and manually annotated and corrected. The chemical translation service is an open tool available to the metabolomics community, which offers highly curated mappings between identifiers from databases that are important for metabolomics [13].

This chapter presents approaches for the calculation of physicochemical parameters, molecular formulae from exact masses, generation of molecular structures, mapping of different database identifiers and their conversion, as well as mapping of metabolites onto pathways. Much emphasis is laid on (semi)automatic processing using command line tools and short R scripts, based on previously published packages. Simple examples show the possibilities and are also easy to use even for beginners.

2 Materials

All presented calculations and scripts were performed on a Windows 7 or 10 64bit machine using R 3.2.5 within RStudio Version 0.99.903, Java 1.8.0_111 64bit, and ChemAxon JChem Suite 16.11.28.0 64 bit. The Parallel Molecule Generator was downloaded from SourceForge (https://sourceforge.net/projects/openmg/). All R libraries were downloaded from the Comprehensive R Archive Network.

2.1 Usage of Command Line Tools

This chapter uses two different types of command line tools. Usage requires a few simple commands to handle the command line. The command line shell can be called in windows by using the windows key and typing cmd and pressing the enter button. This should open a command line in windows similar to the one shown in Fig. 1. This chapter will solely focus on the Windows version of commands.

First, if installed correctly, the ChemAxon tools should be available as environmental variables and can be called by simply typing *cxcalc* or *molconvert*. This can be tested by simply typing *cxcalc*. If the following message appears, the environmental variables are not correctly set: "cxcalc is not recognized as an internal or external command, operable program or batch file." If this is the case, please check the ChemAxon online manual on how to integrate manually the environmental variables. The tools can be also called directly from their installation folder, usually C:\Program Files\ChemAxon\JChemSuite\bin.

The second possibility used in this chapter is to call java packages in a .jar file to perform calculations. For this, Java has to be installed correctly and integrated in the environmental variables. .jar

A

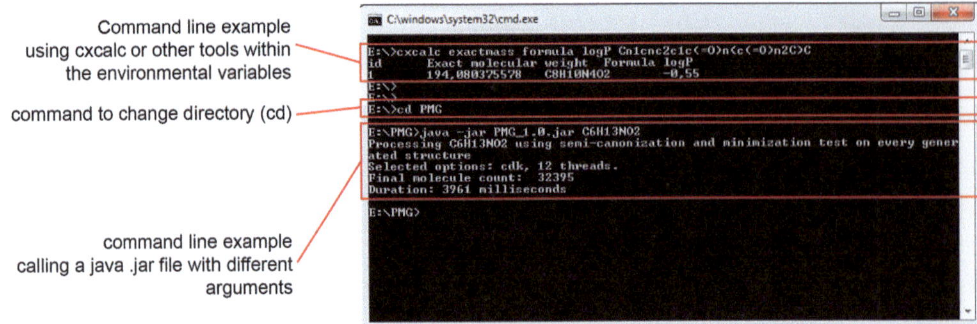

B

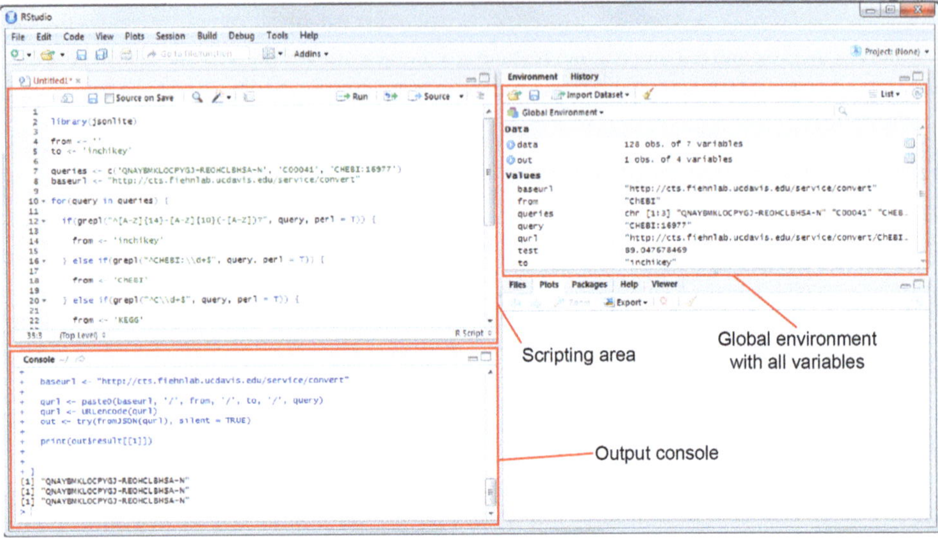

Fig. 1 (**a**) Example for the use of command line tools on a Windows system. The first lines demonstrate the use of the *cxcalc* command from the ChemAxon JChem Suite for calculation of an exact mass, molecular formula, and logP. The following lines show how to call the Parallel Molecule Generator java file to calculate all possible structures for the molecular formula of leucine. (**b**) Screenshot of RStudio showing all basic windows. In the scripting area, all commands can be collected in order to allow repetitive analysis without the need to recall all commands. The global environment contains all currently active variables and allows their inspection for type and size, while the output console allows to run single commands and prints all outputs

files are executed by calling it with the command java –jar XXX.jar args, whereby XXX.jar refers to the .jar file to be used and args is replaced by the argument that shall be passed to the program. An example is shown in the lower part of Fig. 1. An important command to remember is how to change between directories; this is done by using the cd command, e.g., cd <pathToDirectory> changes into the respective directory, while cd .. changes to next higher/lower level in the directories, e.g., from C:\Users\TestUser\ to C:\Users.

2.2 R Scripts

All scripts can be used directly in R or more comfortably with an IDE like RStudio, which allows more convenient work without the need to remember all variables, since they are displayed in a separate window and also allow manual inspection (Fig. 1b). The presented code snippets work with different packages which are not part of the R base distribution and have to be therefore installed manually. Specific packages are jsonlite, XML, rcdk, RCurl, curl, rcdk, OrgMassSpecR, and CHNOSZ with their dependencies. The following script will install all needed packages from CRAN ()https://cran.r-project.org/.

```
#list of packages needed

packages <- c("jsonlite", "XML", "rcdk",
"RCurl", "curl", "OrgMassSpecR", "CHNOSZ")

install.packages(packages, dependencies = T)
```

For the work with R scripts, a basic understanding of R functionalities and data structures is needed (*see* **Note 1**). Most results in the example scripts are returned as a list of specific datatypes, e.g., formula generation with rcdk returns a list of the formal class cdkFormula. Positions in this list are accessed via [i] or [[i]], where i is the index of interest and specific information like the string containing the molecular formula are stored in slots and can be accessed via the @ operator.

```
#get formula string
formulae[[1]]@string
```

Another example is the use of JavaScript Object Notation, short JSON, for transfer of structured data transfer, e.g., from web services. This data format is used by many REST web services for result representation. One example is the Chemical Translation Service offered by the Fiehn Lab. A specific JSON result could look like this.

```
[
 {
 "fromIdentifier": "inchikey",
 "searchTerm": "QNAYBMKLOCPYGJ-REOHCLBHSA-N",
 "toIdentifier": "Chemical Name",
 "result":
 [
 "L-Alanine",
 "L-2-Aminopropanoic acid"
 ]
 }
]
```

The package jsonlite or the *content()* function from the httr package allows a direct conversion of the results in a JSON string to specific datatypes for a more convenient access, which is described in a later section of this chapter.

3   Methods

3.1   Calculation of Exact Mass, logD, and Others Using ChemAxon

Different physicochemical properties are often employed in data analysis of metabolomics data, with exact mass being the most important in mass spectrometry-based metabolomics. The water/octanol partition coefficient logP is often used as a value of molecule polarity and proxy for retention in LC-MS and can help in rejection of false-positive annotations [3].

One package often employed for this task offered by the company ChemAxon is called JChem and includes several command line utilities based on their proprietary Java tools. These command line tools can be used for different purposes. The most employed command for the needs in metabolomics is *cxcalc* used for the calculation of different physicochemical properties. The second most important command is *molconvert*, which allows the interconversion of different chemical structure formats. Lastly, for commands not available via cxcalc directly, *evaluate* can be used. Examples below explain the different use of these commands and how they can be combined in useful blocks (*see* **Notes 2** and **3**).

1. A file containing several structures for testing can be prepared using Microsoft Excel or a text editor. The header should contain #SMILES and Name. The # is important since it labels the first line as header. Structures and names should be separated by a tab. Following entries are used in this paragraph.

   ```
   #SMILES Name
   Cn1cnc2c1c(=O)n(c(=O)n2C)C Caffeine
   CC(C)C[C@@H](C(=O)O)N L-Leucine
   CC[C@H](C)[C@@H](C(=O)O)N L-Isoleucine
   C([C@@H]1[C@H]([C@@H]([C@H]([C@@H](O1)O)O)O)
   O)O D-Glucose
   c1cc(ccc1c2c(c(=O)c3c(cc(cc3o2)O)O)O)O
   Kaempferol
   ```

2. The command *cxcalc* is used for calculating different properties of molecules, whereby the most important are exact mass, formula, and logP. Following command can be used for the simple calculation of all three properties for a single molecule directly in the command line.

   ```
   cxcalc exactmass formula logP Cn1cnc2c1c(=O)
   n(c(=O)n2C)C
   ```

   This function delivers an output on the console with following fields separated by a tab: id, exact molecular weight, formula, logP.

   ```
   id Exact molecular weight Formula logP
   1 194,080375578 C8H10N4O2 -0,55
   ```

   However, if you want to store the result in a file, you have to add the following tag –o<yourPathToResultFile>, e.g.,

```
cxcalc -o E:\cxCalcResult.txt exactmass for-
mula logP
  Cn1cnc2c1c(=O)n(c(=O)n2C)C
```

In the similar way, all structures stored in test.smiles can be converted and stored in a file.

```
cxcalc -o E:\cxCalcResult.txt exactmass for-
mula logP E:\inputSmiles.txt
```

3. Using the commands above, an identifier is reported instead of the original SMILES string. This could be overcome by using the molconvert function and conversion to .sdf format before calculation. The .sdf format allows to store properties in a tag for each molecule. Afterwards the .sdf file can be back converted to SMILES together with the tags. In order not to run three single commands but one, the different commands can be combined by using the pipe | command, which works on both Linux and Windows. The part -T "*" tells the molconvert command to store all calculated properties in tags.

```
molconvert sdf Cn1cnc2c1c(=O)n(c(=O)n2C)C
| cxcalc -S exactmass formula | molconvert
smiles -T "*"

#SMILES name EXACTMASS FORMULA

Cn1cnc2n(C)c(=O)n(C)c(=O)c12 194,080375578
C8H10N4O2
```

This creates the following output, which can be stored again in a file using the –o tag. The {smiles} behind the input file path defines the input as SMILES, which is necessary for molconvert if data is read from a text file.

```
molconvert sdf E:\inputSmiles.txt{smiles} |
cxcalc -S exactmass formula logP | molcon-
vert smiles -T "*" -o E:\cxCalcResult.txt
```

4. SMILES is a proprietary representation of chemical structures and is not open, which lead to several implementations including an open version. Several SMILES can represent the same molecule. In contrast to this, the InChI representation was from the beginning of an open project, which leads to one implementation. Therefore, InChI representations are unique and should be used for the reporting of molecular structures. One major drawback is that InChIs are even harder to read by humans than SMILES. For very efficient reporting, so-called InChI keys, a hashed version of InChIs are used. Both can be generated with the command *evaluate -e molString('inchikey')* or *evaluate -e molString('inchi:AuxNone')* (note: AuxNone suppresses the ChemAxon auxiliary information). The command below first converts all SMILES stored in test.smiles to . sdf format and adds InChI and InChI keys with the tags INCHI

and INCHIKEY. Then exact mass and formula are calculated and finally everything is back converted to SMILES representation with all additional fields and stored in a tab-separated text file.

```
molconvert sdf E:\inputSmiles.
txt{smiles} | evaluate -e "molString('in
chi:AuxNone')" -S -t INCHI | evaluate -e
"molString('inchikey')" -S -t INCHIKEY | cx-
calc -S exactmass formula logP | molconvert
smiles -T "*" -o E:\cxCalcResult.txt
```

3.2 Calculation of Exact Mass, logD, and Others Using RCDK

For people who would not like to use ChemAxon or are not allowed to use it, the Chemistry Development Kit (CDK) is a rich open-source alternative [14, 15]. However, it requires knowledge on programming in Java, and no command line tools are available. The package rcdk provides all CDK functionalities in R and can be accessed easily. An important point is the correct installation of the rJava package which is used to interface java functionalities with R (see **Notes 4** and **5**).

1. First the required library is loaded.

```
#load required libraries
library(rcdk)
```

2. Next, a molecule to work with is generated by parsing a SMILES string, in this case caffeine.

```
#parse smiles
smile <- 'Cn1cnc2c1c(=O)n(c(=O)n2C)C'
mol <- parse.smiles(smile)[[1]]
```

3. Before the molecule can be passed to the functions for the calculation of exact mass and logP, it has to be prepared. This includes detection of aromaticity, type of atoms, and configuration of isotopes.

```
#prepare molecule
do.aromaticity(mol)
do.typing(mol)
do.isotopes(mol)
```

4. After the molecule has been prepared, the *get.exact.mass()*, *get.alogp()*, *get.xlogp()*, *get.tpsa()*, and *get.volume()* functions can be used for calculation of exact mass, logP, total polar surface area, and the molecular volume.

```
#do the calculation
get.exact.mass(mol)
get.alogp(mol)
get.xlogp(mol)
get.tpsa(mol)
get.volume(mol)
```

5. Further descriptors can be calculated with the rcdk package and used for different purposes, e.g., QSRR modeling [16]. Similar to the example above, the molecules have to be parsed and prepared. Molecular descriptors are grouped into different categories, which can be retrieved with the *get.desc.categories()* function. From each category, the descriptor names are available via the *get.desc.names()* function.

```
#get all possible descriptor categories
descriptorCategories <- get.desc.catego-
ries()
descriptorCategories
#get descriptor names of all topological de-
scriptors
descriptorNames <- get.desc.
names(descriptorCategories[3])
descriptorNames
```

6. These names can be used then to tell rcdk which descriptors have to be calculated. Calculation is carried out using the *eval.desc()* function. If all possible descriptors should be calculated, "all" can be used as argument instead of specific descriptors names. The first example calculates only topological descriptors, whereas the second calculates all descriptors. The type of the result is in both cases a list.

```
#get descriptor names of all topological de-
scriptors
descriptorNames <- get.desc.
names(descriptorCategories[3])
descriptorNames
#do the calculation
topologicalDescriptors <- eval.desc(mol, de-
scriptorNames)
topologicalDescriptors
#calculate all descriptors
allDescriptors <- eval.desc(mol, get.desc.
names("all"))
allDescriptors
```

3.3 Formula and Isotope Pattern Calculation Using RCDK

rcdk offers possibilities useful for mass spectrometry-based metabolomics, e.g., calculation of molecular formulas from masses, validation of molecular formulae, or calculation of isotope patterns. A usual first step in metabolite identification is the calculation of molecular formula from a measured exact mass and the comparison of theoretical and measured isotope patterns. Different software tools exist for this purpose. The formula calculation and validation is based on the seven golden rules proposed by Kind and Fiehn [5].

The following examples show how rcdk can be used for the calculation of molecular formulae from a given exact mass.

1. First, the required library is loaded. In this first part, only rcdk is required; however, the packages OrgMassSpecR and CHNOSZ provide useful functionalities used later.

```
#load required libraries
library(rcdk)
library(OrgMassSpecR)
library(CHNOSZ)
```

2. The R script starts from a given isotopic pattern measured ([Glucose + Na]+). First, the monoisotopic mass is located (in this case lowest *m/z* value) and is used as input for the *generate. formula()* function of the rcdk package. This function has several important parameters to set; first the search window has to be set correctly; otherwise, the search for potential sum formulas is too exhaustive. Second, as stated in the seven golden rules, the number of elements has to be restricted. Since the exact mass used as input corresponds to a [M+Na] + adduct, sodium has to be included; additionally, the parameter charge has to set correct, in this case to +1.

```
#get measured isotope pattern ([Glucose +
Na]+)
isoPatternMeasured <- data.frame(mz =
c(203.052609, 204.056051, 205.057227,
206.060394, 207.061845),
int = c(100.000, 6.856, 1.433, 0.087,
0.009))
#get monoisotopic mass
exactmass <- isoPatternMeasured$mz[1]
#calculate all possible formulae
formulae <- generate.formula(exactmass, win-
dow = 0.001,
elements = list(c("C",0,10),c("H",0,50),
c("N",0,5),c("O",0,50),
c("Na",0,1)),
validation = T,
charge = 1)
```

3. In the first case, only one valid chemical formula is produced and returned as a list of cdkFormula objects. Different values can be accessed via the @ operator.

```
#get formula string
formulae[[1]]@string
#get charge
formulae[[1]]@charge
#get mass
formulae[[1]]@mass
```

4. If more than one result is produced, the different results can be accessed by iterating through the returned list. The following example uses the isotope pattern of glutathione as input. Glutathione contains an S atom and therefore has a very specific M+2 pattern. A resolution of 50,000 was used, and glutathione was detected as [M+H] + adduct.

```
#get measured isotope pattern ([Glutathione
+ H]+)
isoPatternMeasured <- data.frame(mz =
c(308.091083, 309.089103, 309.094509,
310.087005, 310.096054, 311.090415,
311.098745),
int = c(100.000, 1.886, 11.251, 4.606,
1.896, 0.524, 0.168))
#get monoisotopic mass
exactmass <- isoPatternMeasured$mz[1]
```

5. Similar to the previous example, the *generate.formula()* function is employed to produce potential formulae. In this case, it is important to consider sulfur in the calculation.

```
#calculate all possible formulae
formulae <- generate.formula(exactmass, win-
dow = 0.001,
elements = list(c("C",0,20),c("H",0,50),
c("N",0,5),c("O",0,50),
c("S",0,5)),
validation = T,
charge = 1)
```

6. The list that is returned contains two different possible molecular formulae. A for loop allows to iterate through this list and access single cdkFormula objects.

```
for(formula in formulae) {
print(formula)
}
```

7. In the next step, the *get.isotopes.pattern()* function is used to simulate an isotopic pattern. The two parameters for this function are a valid cdkFormula object and the minimal abundance of an isotope. At the current stage, the isotope pattern prediction in rcdk does not take into account the charge. Therefore, masses of the predicted isotope pattern have to be corrected for the mass of an electron.

```
#simulate isotopic pattern
isoPatternCalculated <- get.isotopes.
pattern(formula, minAbund = 0.001)
#check for neutral???
#rcdk currently does not support charged
formulas in isotope pattern
```

```
#correction for the mass of a electron need-
ed
isoPatternCalculated[,1] <- isoPatternCalcu-
lated[,1] - 5.48579909070 * 10^-4
```

8. These simulated isotopic patterns have to be compared with the measured for confirmation. Different metrics are available for this. In this example, a simple dot product is used for calculating similarity, offered by the *SpectrumSimilarity()* function of the OrgMassSpecR package, which returns the cosine similarity value between two mass spectra. Since comparison of all different predicted isotope pattern with the measured will yield scores close to 0.9, for a better comparability dissimilarity is used, which is calculated by the following formula: (1—similarity) * 1000. The smaller this value is, the better two isotopic patterns match. Together with the calculation, the function produces a mirror plot of the two (measured and predicted) isotope patterns. The final dissimilarity score is printed but can be also stored in a list or a data frame.

```
#iterate through result list
for(formula in formulae) {
 print(formula)
 #simulate isotopic pattern
 isoPatternCalculated <- get.isotopes.
pattern(formula, minAbund = 0.001)
#check for neutral???
 #rcdk currently does not support charged
formulas in isotope pattern
 #correction for the mass of a electron
needed
 isoPatternCalculated[,1] <- isoPatternCal-
culated[,1] - 5.48579909070 * 10^-4
 #compare measurend and calculated isotopic
pattern
 similarity <- SpectrumSimilarity(isoPattern
Measured, isoPatternCalculated,
 t = 0.001, b = 0.01,
 top.label = "measured isotope pattern",
 bottom.label = "simulated isotope pattern",
 xlim = c(305, 315))
 #label peaks (aesthetics for plotting)
 plot.window(xlim = c(305,315), ylim = c(-
150, 150))
 text(310, 50, paste0("dissimilarity = ",
round((1-similarity)*1000, 2)), cex = 0.75)
 mtext(formula@string, line = 1)
 print((1-similarity)*1000)
}
```

9. Lastly, if further filtering of formulae should be performed, the formula string can be casted into a list with elements and their respective atom counts using the *makeup()* function from the CHNOSZ package.

```
#isolate formula string
formulaString <- formulae[[1]]@string
#parse into named list
elementList <- makeup(formulaString)
elementList #named vector
#get number of carbon atoms
elementList["C"]
```

3.4 Molecular Structure Generation

One major corner stone in identification of unknown metabolites is to calculate potential formulae for a given exact mass as described above. However, this is only the first step. Behind a single sum formula, several isomeric structures can be found. Generation of potential lead structures is therefore a key issue. An open-source tool for structure generation from molecular formulae has been published by Peironcely et al. called Open Molecule Generator (OMG) [17]. A command line tool or a GUI is available from https://sourceforge.net/projects/openmg/, including the Parallel Molecule Generator, which takes advantage of multicore capabilities. A simple example shows how this tool can be used, whereby advanced filtering of structures is performed utilizing ChemAxon tools.

1. The first example calculates all possible structures for $C_6H_{13}NO_2$, the sum formula of L-Leucine, and stores the results in a .sdf file

```
E:\PMG>java -jar PMG_1.0.jar C6H13NO2 -o
leucinePMGExample.sdf
```

2. This generates 32,395 possible molecules. PMG allows valences of 4 for carbon; 2 for oxygen; 3 or 5 for nitrogen; 2, 4, or 6 for sulfur; and 3 or 5 for phosphor. Therefore, the result list may also include weird-looking structures. These results can be cleaned by valence checks, e.g., *cxcalc* offers the possibility to clean the data using Markush enumerations. This filters the molecules based on valences usually found in organic molecules, e.g., 3 for nitrogen.

```
cxcalc -o leucinePMGExampleEnum.sdf enumera-
tions -v -f sdf leucinePMGExample.sdf
```

With this filtering, still 24,843 possible molecules remain, which indicates that further possibilities to restrict the possible chemical space are needed (Fig. 2).

3. The generated, very exhaustive list of possible molecules can be further narrowed down by using the –goodlist and –badlist options, which allows to specify substructures to be included or excluded. Note that the Chemistry Development Kit (cdk)

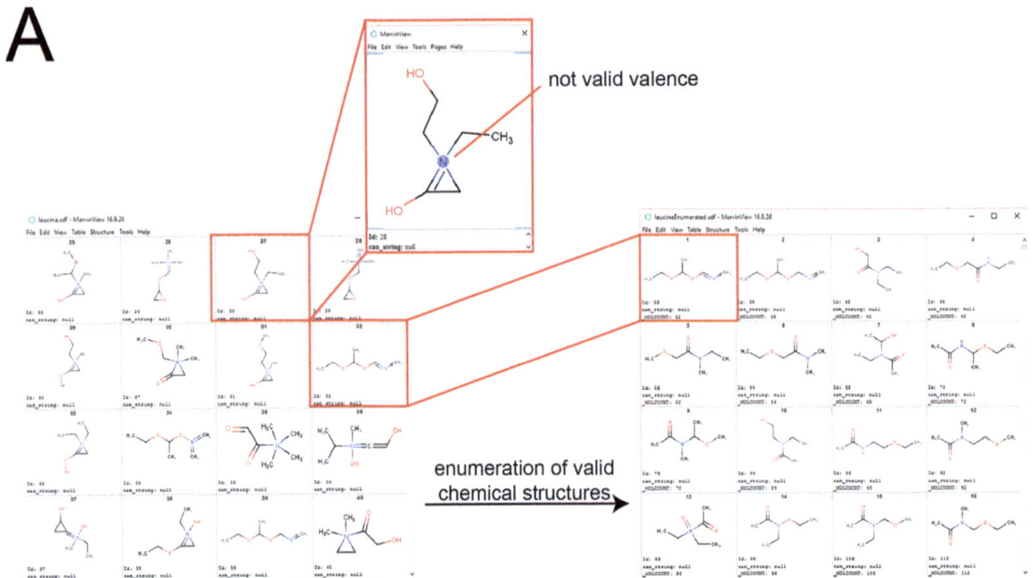

Fig. 2 (a) Enumeration of chemical structures using cxcalc from the JChem tools allows to remove chemical invalid structures, e.g., containing nitrogen atoms of valence 5. In this particular example, all results from structures generated by PMG containing a valence 5 nitrogen were removed, reducing possible chemical structures from 32,395 to 24,843 candidates

has to be installed to make use of these options. For this lists of potential substructures identified by a MS/MS experiment can be included, e.g., a neutral loss of 44 can indicates the presence of a carboxylic acid. In this case, the –goodlist option would include a carboxylic acid. For the leucine example, an amino acid moiety as good list fragment for molecular structure generation is included, and the command is executed.

```
java -jar PMG_1.0.jar C6H13NO2 -goodlist
leucineFrag.sdf -o leucineGoodList.sdf
```

4. These results contain again molecules with nitrogen of valence 5 and can be filtered using the enumeration option as described above.

```
cxcalc -o leucinePMGExampleGoodListEnum.sdf
enumerations -v -f sdf leucinePMGExample-
GoodList.sdf
```

From 56 possible formulas containing an amino acid group, 47 remain after the enumeration of valences. These structures can be used as input for in silico fragmentation tools like MetFrag or retention time prediction [18, 19].

3.5 Identifier Conversion using the Chemical Translation Service

The most unique identifier of a molecule is its structure. However, it is hardly human readable, especially for larger molecules. Still, its uniqueness makes it number one for reporting identified metabolites. SMILES and InChIs are quite long; however, the InChI keys represent an elegant way for structure reporting and can be used in the Chemical Translation Service (CTS), offered by the Fiehn Lab for retrieval of ids of different databases. The most used databases in the field of metabolomics are the Kyoto Encyclopedia of Genes and Genomes, Human Metabolome Database, Lipid Maps, and MetaCyc [20–26]. CTS is available under http://cts.fiehnlab.ucdavis.edu/ and offers single and batch id conversion [13]. However, this is limited to maximum 100 identifiers in one batch. If larger amounts have to be processed, conversion has to be repeated several times. Additionally, if CTS should be integrated into a workflow or several hundreds of identifiers need to be processed, the offered web services are good alternatives. These can be used to convert different database identifiers into each other or search for chemical names from InChI keys. The following examples show that this web service can be invoked in the R environment (*see* **Note 6**).

1. First, the two required libraries for the following examples are loaded.

```
#load required libraries
library(httr)
library(jsonlite)
```

2. The first example uses the *GET()* function from the httr package to retrieve data via a get request from the constructed query URL. Details on the construction of different query URLs can be found on the CTS webpage (http://cts.fiehnlab.ucdavis.edu/moreServices/index). First the identifier from and to which should be converted is defined together with a query.

```
#define from and to identifier
from <- 'inchikey'
to <- 'Chemical Name'
#generate query
queryString <- "QNAYBMKLOCPYGJ-REOHCLBHSA-N"
```

3. Next, a query URL is constructed from the basic URL for conversion, the identifiers that should be employed, and the query itself. This string has to be encoded as URL, since its specific character is not supported in URLs (e.g., blanks are represented as %20). The *URLencode()* function converts specific characters to representation used in URL. Finally, the query is executed using *GET()*. This returns all details of the query including the result, which can be accessed with the *content()* function from the same package, either processed as text, which returns a JSON string, or parsed, which returns a list containing the parsed results.

```
#construct url for GET request
baseUrl <- "http://cts.fiehnlab.ucdavis.edu/
service/convert"
```

```
queryUrl <- paste0(baseUrl, "/", from, "/",
to, "/", queryString)
#use httr GET to retrieve results
queryResult <- GET(URLencode(queryUrl))
#access to content of queryResult
result <- content(queryResult, "text")
result f<- content(queryResult, "parsed")
```

4. Easier access to the web service and the final result can be done by using the fromJSON function from the jsonlite package, which directly performs the request and returns a parsed result list.

```
#use jsonlint fromJSON
queryResult <- fromJSON(URLencode(queryUrl))
```

5. The CTS web services do not allow multiple conversions or mixed input identifiers at the moment. The next example shows what to do if you are dealing with a mixed list of identifiers and you would like to produce a list of one common identifier. Each database identifier has its unique structure, e.g., KEGG compound identifiers consist of a C followed by a five-digit number. This can be used to identify the type identifier based on regular expressions. The MIRIAM registry hosted at the European Bioinformatics Institute contains several hundred identifiers and offers preconfigured regular expressions that can be used (http://www.ebi.ac.uk/miriam/main/mdb?section=intro) [27]. In the next example, different queries are tested with regular expressions for their type and the respective from identifier is used.

```
#define from and to identifier
from <- ''
to <- 'inchikey'
#generate queries
queries <- c('QNAYBMKLOCPYGJ-REOHCLBHSA-N',
'C00041', 'CHEBI:16977')
#base url for construction of query url
baseUrl <- "http://cts.fiehnlab.ucdavis.edu/
service/convert"
for(queryString in queries) {
  if(grepl("^[A-Z]{14}-[A-Z]{10}(-[A-Z])?",
queryString, perl = T)) {
  from <- 'inchikey'
  } else if(grepl("^CHEBI:\\d+$", quer-
yString, perl = T)) {
  from <- 'ChEBI'
  } else if(grepl("^C\\d+$", queryString,
perl = T)) {
  from <- 'KEGG'
  } else {
```

```
}
#construct url for request
baseUrl <- "http://cts.fiehnlab.ucdavis.edu/
service/convert"
queryUrl <- paste0(baseUrl, "/", from, "/",
to, "/", queryString)
#use jsonlint fromJSON
queryResult <-
fromJSON(URLencode(queryUrl))
print(queryResult$result)
}
```

3.6 Manual and Automated Mapping on KEGG Pathways

Both two big pathway servers MetaCyc and KEGG offer web APIs to interact with them. The KEGG database utilizes a simple REST web service for interaction. Different functions allow the retrieval of information on pathways, enzymes, and their linked compounds. This allows querying the DB within a workflow. The functionalities include a mapper, which highlights compounds of interest with default or user-defined colors.

1. This is a simple example how to retrieve a .png file manually using the KEGG Search & Color function (http://www. genome.jp/kegg/tool/map_pathway2.html). In this example, all metabolites of the upper glycolysis pathway are colored in red and all of the lower glycolysis pathway are in blue.

C00267 red	C00074 blue
C00103 red	C00036 blue
C00668 red	C00022 blue
C05345 red	C00186 blue
C05378 red	C00024 blue
C00118 red	C00033 blue
C00110 red	
C00236 red	
C00197 red	
C00631 red	

2. These values can be copied and pasted into the field "Enter objects one per line followed by bgcolor, fgcolor:" (compound input in Fig. 3). If pathway maps of a specific organism should be colored, this organism has to be selected under "Search Against" (organism input in Fig. 3). After pressing the "Exec" button, a list of links to the colored pathway maps is available in the browser. The number in brackets behind each pathway name indicates the number of compounds that have been mapped to this specific pathway.

3. Since manual mapping of several hundreds of pathways can be tedious, an automated version would be useful. This can be achieved by a small R script. The first example just performs simple highlighting of compounds on a single pathway map.

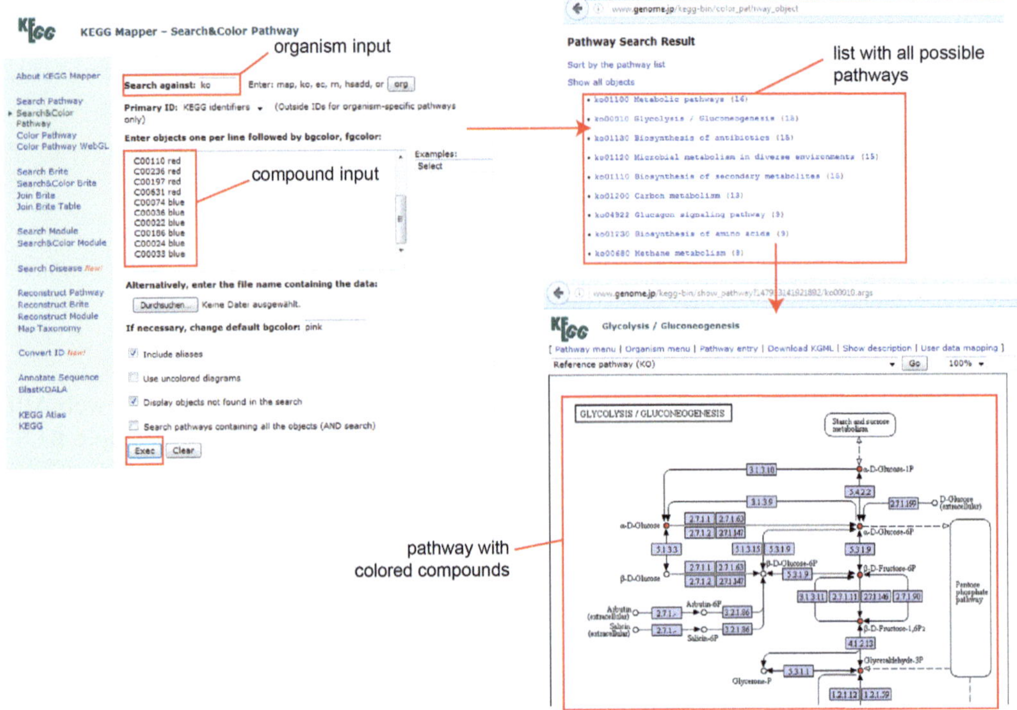

Fig. 3 Manual mapping of compounds is performed with the KEGG Mapper (http://www.genome.jp/kegg/tool/map_pathway2.html). KEGG IDs of compounds of interest are pasted into the text field marked with compound input; optional fore and background color can be defined. An organism can be selected by entering the respective three letter code in the field marked as organism input. After pressing the Exec button, a list with all pathways containing minimum one of the compounds is shown. By clicking on the respective pathway, the map is shown

The link returns a .html page, but a downloadable .png is hidden in the <img> html tag. We utilize the XML package to isolate specifically the html tag that contains the link to the .png file, which is downloaded to the working directory using *curl_download()*. A detail description of the construction of KEGG web link can be found under http://www.kegg.jp/kegg/rest/weblink.html.

```
#generate list with KEGG IDs and bgcolor
compoundList <- list(id = c("C00267",
"C00103", "C00668", "C05345", "C05378",
"C00118", "C00110", "C00236",
"C00197", "C00631", "C00074", "C00036",
"C00022", "C00186", "C00024", "C00033"))
#id of pathway on which metabolites will be
mapped
pathwayId <- "map00010"
baseUrl <- "http://www.kegg.jp/pathway"
#create query url
```

```
queryUrl <- paste0(baseUrl, "/", pathwayId,
"+", paste(compoundList$id, collapse="+"))
#get html content
htmlContent <- getURL(queryUrl)
htmlDoc <- htmlParse(htmlContent)
img <- xpathSApply(htmlDoc, "//img[@
name='pathwayimage']/@src")
#download picture
curl_download(url = paste0("http://www.
kegg.jp", img), destfile = paste0(pathwayId,
".png"))
```

4. Finally, more complicated mapping with self-defined colors can be performed (e.g., defining back and foreground colors). This requires a list of compounds and their respective colors. In the example below, the same coloring as for manual mapping is used. Compound identifiers have to be linked with their colors by a+, and different compound/color pairs are separated by a new line separator "\n." A for loop can be used to iterate through the list of compounds and construct the correct string for the query.

```
#generate list with KEGG IDs and bgcolor
compoundList <- list(id = c("C00267",
"C00103", "C00668", "C05345", "C05378",
"C00118", "C00110", "C00236",
 "C00197", "C00631", "C00074", "C00036",
"C00022", "C00186", "C00024", "C00033"),
 bg = c("red", "red", "red", "red", "red",
"red", "red", "red",
 "red", "red", "blue", "blue", "blue",
"blue", "blue", "blue"))
#create string of compounds and color ac-
cording to KEGG API
compoundString <- ""
for(i in 1:length(compoundList[[1]])) {
 if(i == 1) {
 compoundString <-
paste0(compoundList$id[i],
"+",compoundList$bg[i])
 } else {
 clipboard <- paste0(compoundList$id[i],
"+",compoundList$bg[i])
 compoundString <- paste0(compoundString,
"\n", clipboard)
 }
}
```

5. The query URL is constructed by combining a base URL with a pathway identifier and the string containing the compounds

and colors. *URLencode()* is used to convert the string to its URL representation.

```
#id of pathway on which metabolites will be mapped
pathwayId <- "map00010"
baseUrl <- "http://www.kegg.jp/kegg-bin/show_pathway?map="
#create query url
queryUrl <- URLencode(paste0(baseUrl, pathwayId, "&multi_
query=",compoundString))
#get html content
htmlContent <- getURL(queryUrl)
htmlDoc <- htmlParse(htmlContent)
img <- xpathSApply(htmlDoc, "//img[@name='pathwayimage']/@src")
#download picture
curl_download(url = paste0("http://www.kegg.jp", img), destfile
= paste0(pathwayId, ".png"))
```

---

4 Notes

1. Consult a book or web page on working with R if not familiar, e.g., https://cran.r-project.org/doc/manuals/r-release/R-intro.pdf

2. The command line tools can be invoked into any scripting language which can handle command line tools, e.g., R or Perl. For integration, in Java libraries, all mentioned calculations are available. Please check license from ChemAxon for exact use.

3. The molconvert and cxcalc command line tools need the correct installation of JChem and setting of the respective environmental variables. Please refer to the installation manual.

4. Please check the rJava CRAN package. (Re)Installation of the latest Java version followed by (re)installation of rJava may help with problems.

5. CDK can also use the KNIME workflow environment [28].

6. Once the principle of REST and JSON is clear, the concept can be applied in different web services in metabolomic databases. Lipid Maps offers a REST web service for interaction with their data.

References

1. Benton HP, Wong DM, Trauger SA et al (2008) XCMS2: processing tandem mass spectrometry data for metabolite identification and structural characterization. Anal Chem 80:6382–6389

2. Smith CA, Want EJ, O'Maille G et al (2006) XCMS: processing mass spectrometry data for metabolite profiling using nonlinear peak align-ment, matching, and identification. Anal Chem 78:779–787

3. Müller C, Dietz I, Tziotis D et al (2013) Molecular cartography in acute chlamydia pneumoniae infections—a non-targeted metabolomics approach. Anal Bioanal Chem 405:5119–5131

4. Stanstrup J, Gerlich M, Dragsted LO et al (2013) Metabolite profiling and beyond: approaches for the rapid processing and annotation of human blood serum mass spectrometry data. Anal Bioanal Chem 405(15):5037–5048

5. Kind T, Fiehn O (2007) Seven golden rules for heuristic filtering of molecular formulas obtained by accurate mass spectrometry. BMC Bioinformatics 8:105

6. Tziotis D, Hertkorn N, Schmitt-Kopplin P (2011) Kendrick-analogous network visualisation of ion cyclotron resonance Fourier transform mass spectra: improved options for the assignment of elemental compositions and the classification of organic molecular complexity. Eur J Mass Spectrom 17:415–421

7. Witting M, Lucio M, Tziotis D et al (2015) DI-ICR-FT-MS-based high-throughput deep metabotyping: a case study of the Caenorhabditis Elegans–Pseudomonas Aeruginosa infection model. Anal Bioanal Chem 407:1059–1073

8. Treutler H, Neumann S (2016) Prediction, detection, and validation of isotope clusters in mass spectrometry data. Meta 6:E37

9. Kerber A et al (1998) *MOLGEN 40* Match-communications in mathematical and in computer. Chemistry 37:205–208

10. Peironcely JE et al (2012) OMG: Open Molecule Generator. J Cheminformatics 4:21

11. Jaghoori MM et al (2013) PMG: multi-core Metabolite Identification. Electronic Notes in Theoretical Computer Science 299:53–60

12. Kind T, Scholz M, Fiehn O (2009) How large is the metabolome? A critical analysis of data exchange practices in chemistry. PLoS One 4:e5440

13. Wohlgemuth G et al (2010) The chemical translation service—a web-based tool to improve standardization of metabolomic reports. Bioinformatics 26:2647–2648

14. Steinbeck C et al (2003) The chemistry development kit (CDK): an open-source java library for chemo- and bioinformatics. J Chem Inf Comput Sci 43:493–500

15. Willighagen EL, Mayfield JW, Alvarsson J, Berg A, Carlsson L, Jeliazkova N, Kuhn S, Pluskal T, Rojas-Chertó M, Spjuth O, Torrance G, Evelo CT, Guha R, Steinbeck C (2017) The Chemistry Development Kit (CDK) v2.0: atom typing, depiction, molecular formulas, and substructure searching. J Cheminform 9:33. https://doi.org/10.1186/s13321-017-0220-4

16. Cao M et al (2014) Predicting retention time in hydrophilic interaction liquid chromatography mass spectrometry and its use for peak annotation in metabolomics. Metabolomics:1–11

17. Peironcely JE et al (2012) OMG: Open Molecule Generator. J Cheminformatics 4:1–13

18. Wolf S et al (2010) In silico fragmentation for computer assisted identification of metabolite mass spectra. BMC Bioinformatics 11:148

19. Gerlich M, Neumann S (2013) MetFusion: integration of compound identification strategies. J Mass Spectrom 48:291–298

20. Kanehisa M, Goto S (2000) KEGG: Kyoto encyclopedia of genes and genomes. Nucleic Acids Res 28:27–30

21. Kanehisa M et al (2006) From genomics to chemical genomics: new developments in KEGG. Nucleic Acids Res 34(suppl 1):D354–D357

22. Wishart DS et al (2012) HMDB 3.0—the human metabolome database in 2013. Nucleic Acids Res 41((Database issue)):D801–D807. https://doi.org/10.1093/nar/gks1065

23. Wishart DS et al (2009) HMDB: a knowledge-base for the human metabolome. Nucleic Acids Res 37(Database):D603–D610

24. Sud M et al (2007) LMSD: LIPID MAPS structure database. Nucleic Acids Res 35(suppl 1):D527–D532

25. Caspi R et al (2008) The MetaCyc database of metabolic pathways and enzymes and the BioCyc collection of pathway/genome databases. Nucleic Acids Res 36(suppl 1):D623–D631

26. David S. Wishart, Yannick Djoumbou Feunang, Ana Marcu, An Chi Guo, Kevin Liang, Rosa Vázquez-Fresno, Tanvir Sajed, Daniel Johnson, Carin Li, Naama Karu, Zinat Sayeeda, Elvis Lo, Nazanin Assempour, Mark Berjanskii, Sandeep Singhal, David Arndt, Yonjie Liang, Hasan Badran, Jason Grant, Arnau Serra-Cayuela, Yifeng Liu, Rupa Mandal, Vanessa Neveu, Allison Pon, Craig Knox, Michael Wilson, Claudine Manach, Augustin Scalbert; HMDB 4.0: the human metabolome database for 2018, Nucleic Acids Research, gkx1089, https://doi.org/10.1093/nar/gkx1089

27. Juty N, Le Novère N, Laibe C (2012) Identifiers.org and MIRIAM registry: community resources to provide persistent identification. Nucleic Acids Res 40:D580–D586

28. Beisken S et al (2013) KNIME-CDK: workflow-driven cheminformatics. BMC Bioinformatics 14:257

Part II

Methods

Chapter 5

HILIC-MS/MS Multi-Targeted Method for Metabolomics Applications

Christina Virgiliou, Helen G. Gika, and Georgios A. Theodoridis

Abstract

Metabolomics aims at the identification and quantification of key-end point metabolites, basically polar, in order to study changes in biochemical activities in response to pathophysiological stimuli or genetic modifications. Targeted profiling assays have enjoyed a growing popularity during the last years with LC-MS/MS as a powerful tool for development of such (semi-) quantitative methods for a large number of metabolites. Here we describe a method for absolute quantification of ca. 100 metabolites belonging to key metabolite classes such as sugars, amino acids, nucleotides, organic acids, and amines with a hydrophilic interaction liquid chromatography (HILIC) system comprised of ultra (high) performance liquid chromatography (UHPLC) with detection on a triple-quadrupole mass spectrometer operating in both positive and negative electrospray ionization modes.

Key words HILIC-MS/MS, Metabolic profiling, Targeted metabolomics, Polar analytes

---

1   Introduction

Metabonomics or metabolomics, often described as the holistic metabolic profile of complex matrices such as biological fluids, tissue, and cell extracts, represents with genomics and proteomics the major platforms in system biology [1]. Developments in analytical chemistry and particular advances in separation and spectroscopic technologies made metabolomics a rapidly developed research field over the past few decades [2, 3]. Although mass spectrometry and NMR are the most popular analytical platforms used for metabolomics-based studies, emphasis is lately placed on LC/MS approaches due to a wide range of coverage of metabolites and its high efficiency [4]. Typically MS-studies for metabolite profiling can be categorized as targeted and untargeted/holistic approaches.

Georgios A. Theodoridis et al. (eds.), *Metabolic Profiling: Methods and Protocols*, Methods in Molecular Biology, vol. 1738, https://doi.org/10.1007/978-1-4939-7643-0_5, © Springer Science+Business Media, LLC, part of Springer Nature 2018

Targeted analysis is usually a hypothesis-driven strategy. It focuses mainly on the measurement (identification and quantification) of selected metabolites with known chemical properties, thus sample preparation can be adapted in order to minimize limitations including matrix effect [5]. On the other hand, the holistic approach investigates the whole metabolic complement of the analyzed sample. However, it has been realized that there are a number of issues that need to be resolved for untargeted MS-studies including standardization, robustness, and reproducibility that affect the validity of the results [6–8]. Limitations of MS-methods and current developments on triple-quadrupole instrumentation along with advanced software capabilities have resulted in the development of tailor-made targeted metabolomics methods able to (semi)-quantify tens of analytes of specific interest in a single injection and provide solid, quantitative, and unambiguous data [9–11].

The development of a comprehensive method for targeted metabolomics represents a challenge. The samples of interest often contain highly polar metabolites with different physicochemical properties, coexisting in samples over different concentration ranges. Additional efforts are required for fine-tuning of all analytical parameters in order to find the optimum for most of the analytes measured within the method. According to the literature, a number of multi-analytes methods use hydrophilic interaction liquid chromatography (HILIC) separations for simultaneous analysis of large numbers of polar metabolites [12, 13]. Reversed phase (RP) chromatography is not able to retain polar metabolites, while although the ion-pair chromatography has been reported as the most powerful option ion-pair reagent contaminates the MS system to a major extent [14, 15].

In the present protocol, a procedure for the identification and quantification of ca. 100 metabolites via (HILIC) UPLC-MS/MS method is described. The aim of the method is the identification and quantitation of key-end point metabolites known to exist in biological fluids (serum, plasma, urine, amniotic fluid) in order to study their metabolic profile. Mass spectrometer and chromatographic condition were optimized in order to achieve satisfactory detection, quantification, maximum peak capacity, and retention for as many metabolites as possible in a single run [16, 17]. Absolute quantitation can be performed using the standard addition approach. Matrix effect and recovery can be estimated, and the sample preparation procedure is optimized in order to reduce matrix effects particularly for blood samples.

2   Materials

All solvents used should be of LC/MS analytical grade. Use purified water (18.2 MΩ, at 25 °C). All used standards should be of analytical or higher grade. Stock and working standard solutions should be kept at −20 °C, and solvents containing >50% of water should be replaced after 20 days (*see* **Note 1**).

2.1 Stock and Working Standard Solutions

Classify compounds in different concentration groups: group A, group B, and group C (details in Table 1) (*see* **Notes 2** and **3**).

Stock solutions of the analytes should be prepared in concentrations of (group A) 1000, (group B) 5000, and (group C) 10,000 mg/L in methanol(MeOH)/water, 1:1 (v/v) or water, depending on analyte solubility (*see* **Note 4**). Prepare working standards mixtures from the stock solution by appropriate dilution with acetonitrile(MeCN)/water, 95:5 (v/v).

Calibration standards (9 standards) should be prepared by serial dilution of the highest concentration standard.

For the standard mixture with the highest concentration (std 9): Mix 0.0032 ml of each standard in group A with 0.052 ml of each standard in group B and 0.152 ml of each standard in group C. In order to reach the final volume of 8 ml dilute with 95:5 MeCN:H_2O. Repeat this step depending on the desired final volume of std. 9.

2.2 Liquid Chromatography

Mobile phase A, stock buffer solution: In a 50 ml beaker, weigh 0.631 g ammonium formate and add H_2O (<50 ml) (MilliQ). Sonicate until salt is fully dissolved. Pour the sonicated solvent into a 50 ml volumetric flask and add H_2O MilliQ till the calibration mark. Final ammonium formate buffer concentration: 200 mM.

Mobile phase A, 95:5 MeCN:H_2O, 10 mM ammonium formate: Pour the 50 ml buffer into a clean 1000 ml Schott bottle. Add gradually 950 ml MeCN. Shake and sonicate for homogenization and place in ultrasonic bath for degassing (*see* **Note 5**).

Mobile phase B, stock buffer solution: In a 50 ml beaker, weigh 0.2254 g ammonium formate and add water (<250 ml) (MilliQ). Sonicate until salt is fully dissolved. Pour dissolved buffer into a 250 ml volumetric flask and add H_2O MilliQ till the calibration line. Final ammonium formate buffer concentration: 14.2 mM.

Mobile Phase B, 30:70 MeCN:H_2O, 10 mM ammonium formate: With the use of a clean volumetric cylinder measure 210 ml

Table 1
Group concentration and nominal concentration of each group in standards 1–9 used for the construction of calibration curves

Meatbolites	Group concentration	Meatbolites	Group concentration	Meatbolites	Group concentration	Meatbolites	Group concentration	Standard	Mixture	mg/L
2-Hydroxyisobutyric acid	A	Cytidine	A	Lactose	B	Sorbitol	B	STD 1	A	0.01
2-Hydroxyisovaleric acid	A	Cytosine	A	Leucine	A	Suberic acid	A		B	0.09
2-MethylHippuric acid	A	Dimethylamine	A	Lysine	B	Sucrose	A		C	0.475
3-Methylhistidine	B	Folic acid	A	Malic acid	A	Taurine	A	STD 2	A	0.02
4-Hydroxyphenyllactate	B	Fructose	B	Malonic acid	B	Theobromine	B		B	0.18
α-Ketoglutaric acid	B	Fucose	B	Maltose	B	Thiamine	B		C	0.95
Acetylcarnitine	A	Fumaric acid	A	Mannitol	B	Threonine	B	STD 3	A	0.08
Adenine	A	γ-Aminobutyric acid	A	Methionine	A	Thymidine	A		B	0.72
Adenosine	A	Galactosamine	A	Methylamine	A	Thymine	B		C	3.8
Adipic acid	A	Galactose	B	Monoisoamylamine	A	Trimethylamine	A	STD 4	A	0.2
Alanine	B	Glucose	C	N-AcetylAspartate	A	Trimethylamine-n-oxide	A		B	1.8
Arabitol	B	Glutamic acid	B	Nicotinamide	B	Tryptamine	A		C	9.5
Arginine	A	Glutamine	A	Nicotinic acid	B	Tryptophan	A	STD 5	A	0.5
Ascorbic acid	B	Glycine	A	Ornithine	B	Tyrosine	B		B	4.5
Asparagine	B	Guanine	B	Pantothenic acid	B	Uracil	A		C	23.75
Aspartic acid	B	Hippuric acid	B	Phenylalanine	B	Uric acid	C	STD 6	A	1
Benzoic acid	A	Histamine	A	Picolinic acid	A	Uridine	A		B	9
Betaine	A	Homocysteine	B	Proline	B	Valine	B		C	47.5
Biotin	A	Hypotaurine	A	Putrescine	A	Vitamin B12	A	STD 7	A	1.2
Cadaverine	A	Hypoxanthine	A	Pyridoxine	A	Xanthine	A		B	10.8
Caffeine	A	Inosine	A	Pyroglutamic acid	B	Xanthurenic acid	B		C	57
Choline	A	Inositol	B	Pyruvic acid	B	Xylitol	B	STD 8	A	1.5
Cotinine	A	Isoleucine	A	Ribose	B	Xylose	B		B	13.5
Creatine	A	Itaconic acid	A	Riboflavine	A				C	71.25
Creatinine	A	Kynurenic acid	A	Sarcosine	A			STD 9	A	2
Cystine	B	Lactic acid	C	Serine	B				B	18
									C	95

buffer and pour into a 500 ml Schott bottle. Measure 90 ml MeCN and pour it gradually to the buffer. Shake and sonicate for homogenization and degassing.

Wash, weak solvent: Measure 420 ml water and pour into a clean 1 L Schott bottle, measure 180 ml of methanol, and add to water. Finally add 0.1% formic acid (600 µl).

Purge, strong solvent: Measure 380 ml acetonitrile and pour into a clean 1 L Schott bottle, measure 20 ml of methanol and add to water. Finally add 0.1% Formic acid (400 µl).

Seal wash solvent: Measure 900 ml of water and pour into a 1 L Schott bottle, measure 100 ml of MeCN, and add to water. Mix thoroughly and degas shortly in ultrasonic bath.

Equipment and column: A Waters Acquity *H* Class UPLC system coupled to Xevo TQD MS-spectrometer under control of MassLynx 4.1 is used for the present protocol. In cases where the sample volume is limited, total recovery vials (or similar type) with pre-slit silicone septa screw caps 9 mm were used. Chromatography is performed on an Acquity BEH Amide Column (2.1 mm i.d. × 150 mm, 1.7 µm) protected by an Acquity UPLC Van-Guard pre-column.

3 Methods

Methods for LC-MS/MS are prepared and stored in the centrally shared hard drive.

3.1 UPLC Method

With regard to UPLC conditions, the parameters for binary solvent manager are as follows: flow rate is kept constant throughout the whole analysis at 0.5 ml/min and the following gradient is programmed: 4 min isocratic step at 100% A, then rising to 40% B linearly over the next 21 min and finally reaching 85% B over 5 min. The column is equilibrated for 10 min in the initial conditions. Regarding the sample manager, the flow through needle system is applied in the present protocol. Injection volume is set at 5 µl and sample temperature at 6 °C. Injection system is subjected to two washing cycles with a strong solvent and a weak solvent prior to injection and one cycle of 6 s with the strong solvent for post-wash. Column temperature is set to 40 °C.

3.2 MS Method

In order to edit the MS method, find the optimum parameters for the MRM transition of each metabolite. For this protocol, manual optimization is performed in order to find precursor and product ions and the optimum cone voltage and collision energy. With regard to capillary voltage, the best possible for most metabolites detected in positive and negative ionization mode is applied. All the other parameters were set according to the tune page and the linked calibration file.

Apply multiple reaction monitoring (MRM) mode for the detection and quantification of all the compounds. Operate electrospray ionization at polarity switching mode. Set capillary voltage at +3.5 kV or −3.5 kV, block and desolvation temperatures at 150 °C and 350 °C, respectively. Set desolvation gas flow rate at 650 L/h and cone gas at 50 L/h. Optimum cone voltage and collision energy for each analyte after direct infusion and optimum time window and dwell times are presented in Table 2.

System start-up and pretests.

Follow each step here to ensure high quality of your data. For the system start-up, the system is prepared for analysis and checked for system suitability.

1. Place beakers with mobile phase A and B, wash, purge, and seal wash solvents to the corresponding channels and in the two left channels, install neat acetonitrile and water.

2. Prime eluents for 3 min each.

3. Connect the BEH Amide column.

4. Set the column temperature to 40 °C and the temperature of sample manager to 6 °C.

5. Flush the column with 95% MeCN and 5% water for 15 min at flow of 0.2 ml/min.

6. Increase flow rate gradually to 0.5 ml/min.

7. Flush column until equilibration (psi delta <20) (*see* **Note 6**),

8. Load UPLC method.

For MS, load the appropriate Acquity DB file with the extension .ipr and the most recent calibration file.

3.3 Sample Preparation

Extraction of samples may vary between different matrices. So far the described method or its variants is tested with serum, urine, amniotic fluid, intra-/extracellular content, feces, and various types of animal tissue but also in foods such as honey, muscle tissue, and flour. The sample preparation procedure for blood serum samples is presented below.

1. Allow samples to thaw at room temperature.

2. Mix 50 μl of sample with 130 μl MeCN, 10 μl H_2O, and 10 μl MeOH in an eppendorf vial of 1 ml by the use of variable volume pipette 20–200 μl (*see* **Note 7**).

Table 2
Analytes that can be detected and monitored with the HILIC-MS/MS method

A/A	Metabolites	Formula	Monoisotopic mass	Precursor ion	Product ion	Cone voltage (V)	Collision energy (V)	Polarity	Rt (min)	Molecular weight	Dwell time
1	2-Hydroxyisobutyric acid	$C_4H_8O_3$	104.04	103	57	30	10	−	8.0	104.10	0.005
2	2-Hydroxyisovaleric acid	$C_5H_{10}O_3$	118.06	117	71	30	12	−	6.0	118.13	0.005
3	2-MethylHippuric acid	$C_{10}H_{11}NO_3$	193.20	192	148	35	12	−	8.2	193.20	0.005
4	3-Methylhistidine	$C_7H_{11}N_3O_2$	169.09	170	109	30	10	+	19.0	169.18	0.003
5	4-Hydroxyphenyllactate	$C_9H_{10}O_4$	182.05	181	63	33	12	−	10.0	182.17	0.02
6	a-Ketoglutaric acid	$C_5H_6O_5$	146.01	145	101	20	9	−	16.0	146.11	0.02
7	Acetylcarnitine	$C_9H_{17}NO_4$	203.12	204	85	30	10	+	14.4	203.23	0.003
8	Adenine	$C_5H_5N_5$	135.05	136	119	40	20	+	3.6	135.13	0.003
9	Adenosine	$C_{10}H_{13}N_5O_4$	267.10	268	136	20	15	+	4.4	267.24	0.003
10	Adipic acid	$C_6H_{10}O_4$	146.06	145	101	25	12	−	16.0	146.14	0.005
12	Alanine	$C_3H_7NO_2$	89.05	90	44	20	10	+	16.0	89.09	0.005
13	Arabitol	$C_5H_{12}O_5$	152.07	151	89	25	10	−	9.9	152.14	0.02
14	Arginine	$C_6H_{14}N_4O_2$	174.11	175	70	30	19	+	21.9	174.20	0.005
15	Ascorbic acid	$C_6H_8O_6$	176.03	176	70	20	15	+	3.6	176.12	0.02
16	Asparagine	$C_4H_8N_2O_3$	132.05	133	74	20	14	+	18.2	132.11	0.02
17	Aspartic acid	$C_4H_7NO_4$	133.04	134	74	18	16	+	21.8	133.11	0.02
18	Benzoic acid	$C_7H_6O_2$	122.04	121	77	25	11	−	1.8	122.12	0.032
19	Betaine	$C_5H_{11}NO_2$	117.07	118	59	38	18	+	12.5	117.15	0.005

(continued)

Table 2
(continued)

A/A	Metabolites	Formula	Monoisotopic mass	Precursor ion	Product ion	Cone voltage (V)	Collision energy (V)	Polarity	Rt (min)	Molecular weight	Dwell time
20	Biotin	$C_{10}H_{16}N_2O_3S$	244.09	245	227	25	14	+	8.1	244.31	0.003
21	Cadaverine	$C_5H_{14}N_2$	102.12	103	86	15	8	+	20.4	102.17	0.003
22	Caffeine	$C_8H_{10}N_4O_2$	194.08	195	138	38	18	+	0.9	194.19	0.032
23	Choline	$C_5H_{14}NO$	104.11	104	60	40	22	+	7.0	104.17	0.003
25	Cotinine	$C_{10}H_{12}N_2O$	176.09	117	80	30	20	+	1.1	176.22	0.032
26	Creatine	$C_4H_9N_3O_2$	131.07	132	90	28	10	+	16.3	131.11	0.003
27	Creatinine	$C_4H_7N_3O$	113.06	114	88	30	10	+	4.8	113.11	0.003
29	Cystine	$C_6H_{12}N_2O_4S_2$	240.02	241	152	26	12	+	24.6	240.30	0.02
30	Cytidine	$C_9H_{13}N_3O_5$	243.09	244	112	15	10	−	11.0	243.22	0.005
31	Cytosine	$C_4H_5N_3O$	111.04	112	95	40	14	+	7.5	111.10	0.003
32	Dimethylamine	C_2H_7N	45.06	46	30	30	30	+	8.2	45.08	0.005
33	Folic acid	$C_{19}H_{19}N_7O_6$	441.14	442	295	22	13	+	22.4	441.39	0.005
34	Fructose	$C_6H_{12}O_6$	180.06	181	140	5	8	+	11.8	180.16	0.01
35	Fucose	$C_6H_{12}O_5$	164.06	163	59	22	12	−	7.0	164.16	0.005
36	Fumaric acid	$C_4H_4O_4$	116.01	115	71	25	8	−	20.3	116.07	0.02
37	g-Aminobutyric acid	$C_4H_9NO_2$	103.06	104	69	22	15	+	17.3	103.12	0.005
38	Galactosamine	$C_6H_{13}NO_5$	179.07	180	72	18	18	+	17.8	179.17	0.01
39	Galactose	$C_6H_{12}O_6$	180.06	179	89	15	8	−	11.9	180.15	0.02

A/A	Metabolites	Formula	Monoisotopic mass	Precursor ion	Product ion	Cone voltage (V)	Collision energy (V)	Polarity	Rt (min)	Molecular weight	Dwell time
40	Glucose	$C_6H_{12}O_6$	180.06	179	59	25	16	−	14.6	180.15	0.02
41	Glutamic acid	$C_5H_9NO_4$	147.05	130	84	25	16	+	21.0	147.13	0.02
42	Glutamine	$C_5H_{10}N_2O_3$	146.07	148	84	20	15	+	17.8	146.14	0.005
43	Glycine	$C_2H_5NO_2$	75.03	76	30	35	6	+	17.0	75.06	0.02
44	Guanine	$C_5H_5N_5O$	151.05	152	135	35	17	+	10.0	151.13	0.005
45	Hippuric acid	$C_9H_9NO_3$	179.06	178	134	32	11	−	9.4	179.17	0.005
46	Histamine	$C_5H_9N_3$	111.08	112	95	23	12	+	13.7	11.15	0.003
48	Homocysteine	$C_4H_9NO_2S$	135.03	134	88	10	8	−	16.4	135.19	0.005
49	Hypotaurine	$C_2H_7NO_2S$	109.02	110	92	22	18	+	15.8	109.14	0.015
50	Hypoxanthine	$C_5H_4N_4O$	136.04	137	110	40	18	+	4.8	136.11	0.003
51	Inosine	$C_{10}H_{12}N_4O_5$	268.08	269	137	15	10	+	9.2	268.22	0.005
52	Inositol	$C_6H_{12}O_6$	180.06	181	109	15	10	+	17.6	180.16	0.02
53	Isoleucine	$C_6H_{13}NO_2$	131.09	132	86	25	12	+	13.3	131.17	0.003
54	Itaconic acid	$C_5H_6O_4$	130.03	129	85	20	8	−	14.4	130.09	0.02
55	Kynurenic acid	$C_{10}H_7NO_3$	189.04	190	172	32	12	+	9.8	189.16	0.005
56	Lactic acid	$C_3H_6O_3$	90.03	89	43	30	10	−	11.7	90.08	0.02
57	Lactose	$C_{12}H_{22}O_{11}$	342.12	343	163	10	10	+	18.5	342.30	0.02
58	Leucine	$C_6H_{13}NO_2$	131.09	132	86	20	10	+	13.4	131.17	0.003
59	Lysine	$C_6H_{14}N_2O_2$	146.11	147	84	14	14	+	22.5	146.19	0.01

(continued)

Table 2
(continued)

A/A	Metabolites	Formula	Monoisotopic mass	Precursor ion	Product ion	Cone voltage (V)	Collision energy (V)	Polarity	Rt (min)	Molecular weight	Dwell time
60	Malic acid	$C_4H_6O_5$	134.02	133	115	22	10	−	20.2	134.08	0.02
61	Malonic acid	$C_3H_4O_4$	104.01	103	59	15	9	−	14.4	104.06	0.02
62	Maltose	$C_{12}H_{22}O_{11}$	342.12	341	161	25	8	−	18.1	342.30	0.02
63	Mannitol	$C_6H_{14}O_6$	182.08	183	69	15	11	+	13.5	182.17	0.02
65	Methionine	$C_5H_{11}NO_2S$	149.05	150	104	22	9	+	14.0	149.21	0.005
66	Methylamine	CH_5N	31.04	32	32	25	3	+	10.0	31.05	0.005
67	Monoisoamylamine	$C_5H_{13}N$	87.10	88	43	18	11	+	4.6	87.10	0.003
68	N-AcetylAspartate	$C_6H_9NO_5$	175.04	176	134	18	10	+	20.1	175.14	0.005
69	Nicotinamide	$C_6H_6N_2O$	122.05	123	96	40	15	+	1.1	122.12	0.032
70	Nicotinic acid	$C_6H_5NO_2$	123.03	124	80	38	18	+	10.5	123.10	0.005
71	Ornithine	$C_5H_{12}N_2O_2$	132.09	133	70	45	10	+	22.7	132.16	0.02
72	Pantothenic acid	$C_9H_{16}NO_5$	218.10	220	90	25	13	+	12.2	219.23	0.005
73	Phenylalanine	$C_9H_{11}NO_2$	165.08	166	120	22	12	+	12.6	165.19	0.003
74	Picolinic acid	$C_6H_5NO_2$	123.03	124	106	25	10	+	28.9	123.10	0.005
75	Proline	$C_5H_9NO_2$	115.06	116	70	20	20	+	14.5	115.13	0.02
76	Putrescine	$C_4H_{12}N_2$	88.10	89	72	15	8	+	21.0	88.15	0.003
77	Pyridoxine	$C_8H_{11}NO_3$	169.07	170	152	28	12	+	2.0	169.18	0.003
78	Pyroglutamic acid	$C_5H_7NO_3$	129.04	130	84	30	12	+	15.0	129.11	0.005

A/A	Metabolites	Formula	Monoisotopic mass	Precursor ion	Product ion	Cone voltage (V)	Collision energy (V)	Polarity	Rt (min)	Molecular weight	Dwell time
79	Pyruvic acid	$C_3H_4O_3$	88.02	89	48	20	12	+	7.1	88.06	0.02
81	Ribose	$C_5H_{10}O_5$	150.05	149	89	22	10	−	4.3	150.13	0.005
82	Riboflavine	$C_{17}H_2ON_4O_6$	376.14	377	243	38	22	+	9.6	376.36	0.005
83	Sarcosine	$C_3H_7NO_2$	89.05	90	44	20	8	+	15.3	89.09	0.005
84	Serine	$C_3H_7NO_3$	105.04	106	60	20	10	+	17.9	105.09	0.005
85	Sorbitol	$C_6H_{14}O_6$	182.08	181	101	35	10	−	13.5	182.17	0.02
87	Suberic acid	$C_8H_{14}O_4$	174.09	173	111	35	12	−	10.7	174.20	0.005
88	Sucrose	$C_{12}H_{22}O_{11}$	342.12	341	179	40	14	−	16.8	342.29	0.02
89	Taurine	$C_2H_7NO_3S$	125.01	126	108	25	10	+	14.4	125.14	0.005
90	Theobromine	$C_7H_8N_4O_2$	180.06	181	163	35	17	+	1.1	180.16	0.032
91	Thiamine	$C_{12}H_{17}N_4OS$	265.11	265	122	22	12	+	11.8	300.81	0.003
92	Threonine	$C_4H_9NO_3$	119.06	120	74	20	10	+	10.0	119.11	0.005
93	Thymidine	$C_{10}H_{14}N_2O_5$	242.09	243	127	11	9	+	1.6	242.22	0.032
94	Thymine	$C_5H_6N_2O_2$	126.04	127	110	40	19	+	1.2	126.11	0.032
95	Trimethylamine	C_3H_9N	59.07	60	45	31	10	+	5.6	59.11	0.005
96	Trimethylamine-n-oxide	C_3H_9NO	75.07	76	59	28	10	+	13.0	75.10	0.003
97	Tryptamine	$C_{10}H_{12}N_2$	160.10	161	144	15	11	+	4.1	160.21	0.003
98	Tryptophan	$C_{11}H_{12}N_2O_2$	204.09	205	146	20	15	+	12.7	204.22	0.005
99	Tyrosine	$C_9H_{11}NO_3$	181.07	182	136	22	13	+	14.5	181.19	0.005

(continued)

Christina Virgiliou et al.

Table 2
(continued)

A/A	Metabolites	Formula	Monoisotopic mass	Precursor ion	Product ion	Cone voltage (V)	Collision energy (V)	Polarity	Rt (min)	Molecular weight	Dwell time
100	Uracil	$C_4H_4N_2O_2$	112.03	113	70	40	15	+	1.9	112.08	0.02
101	Uric acid	$C_5H_4N_4O_3$	168.03	169	141	35	15	+	16.3	168.11	0.02
102	Uridine	$C_9H_{12}N_2O_6$	244.07	243	110	35	16	−	4.7	244.12	0.005
103	Valine	$C_5H_{11}NO_2$	117.08	118	72	20	10	+	14.4	117.15	0.003
104	Vitamin B12	$C_{63}H_{89}CoN_{14}O_{14}P$	1355.58	678	147	42	27	+	19.4	1335.37	0.005
105	Xanthine	$C_5H_4N_4O_2$	152.03	153	136	33	15	+	7.4	152.11	0.01
106	Xanthurenic acid	$C_{10}H_7NO_4$	205.04	206	188	35	12	+	12.8	205.17	0.02
107	Xylitol	$C_5H_{12}O_5$	152.07	151	89	30	10	−	9.9	152.15	0.02
108	Xylose	$C_5H_{10}O_5$	150.13	149	89	25	7	−	8.0	150.13	0.005

3. Vortex for 10 min and centrifuge at 5480 × g-force for 10 min.

4. Transfer supernatant to an LC-MS with glass insert vial.

5. Immediately transfer to precooled sample manager.

For standard addition approach, sample preparation of spiking samples is as follows (*see* **Note 8**):

1. Mix 50 μl of sample with 100 μl of standard calibration mixture 1 and 50 μl of MeCN in an eppendorf vial of 1 ml by the use of a variable volume pipette 20–200 μl.

2. Vortex for 10 min and centrifuge for 10 min.

3. Transfer supernatant to an LC-MS with glass insert vial.

4. Repeat **steps 1–3** using increased concentration standard calibration mixtures (at least 3 in total).

5. Immediately transfer to precooled sample manager.

In case of external calibration curve transfer standards to LC-MS vial with glass insert (*see* **Notes 9** and **10**).

For quality control sample (QC sample), in order to evaluate systems stability, mix equal volumes of all samples of the dataset and follow the sample preparation procedure.

3.4   UPLC-MS/MS Analysis

The following steps describe how to set up a sample table for data acquisition. The following sample order has been evaluated to be optimal for both throughput and quality controls:

1. Run a gradient without injection in order to evaluate column performance and have a measure of minimum and maximum pressure during analysis (*see* **Note 11**).

2. Perform six replicate injections of a standard mixture in order to pretest systems performance (retention time, signal) and to equilibrate system (*see* **Note 12**).

During analysis with external calibration approach:

1. Run a before-batch calibration curve.

2. Run injections of a QC sample in order to perform matrix equilibration of the system. The number of equilibration injections depends on the analyzed matrix. For cell media and urine five injections may be adequate, for blood and tissue samples higher numbers may be necessary, depending also on the use of the column (new columns need more injections to saturate active sites/equilibrate).

3. Samples are injected in random order in blocks of ten samples.

4. After each block of ten samples, injection of a standard mixture and a QC sample is performed.

5. Repeat **steps 2** and **3** for maximum 100 samples (*see* **Note 13**).

6. Run an after-batch calibration curve.

During analysis with standard addition approach

1. Run injections of a QC sample in order to perform matrix equilibration of the system.

2. Samples followed by their paired spiking samples are injected in order in block of ten samples.

3. After each block of ten samples, injection of a standard mixture and a QC sample is performed.

4. Repeat **steps 2** and **3** for maximum 100 samples.

After finishing sample batch analysis column and the MS have to be cleaned for further use.

1. Flush the column with 50% MeCN and 50% H_2O for 50 min, column temperature 50 °C, flow 0.2 ml/min.

2. Flush the column with 95% MeCN and 5% H_2O for 30 min, column temperature 40 °C, flow rate 0.5 ml/min.

3. Clean MS cone as recommended.

3.5 Data Treatment-Quantification of Metabolites

Quantitation can be performed in both manual and automated ways using vendor software. In the present protocol TargetLynx (Waters) is used. A quantify method must be created before integration or quantification can be performed. Related software from other vendors includes MultiQuant (Sciex), Xcalibur (Thermo Fischer), and MassHunter WorkStation-Quantitative Analysis (Agilent).

TargetLynx data can be saved as .qld files for further manipulation. Complete summary of the results and .qld files (area, response, concentration, S/N, SD, measured concentration, etc.) can be exported as .txt files and open with excel Microsoft program for further treatment.

In case of external calibration curve approach, once you find the optimum method parameters, calibration of standards and quantification can be performed directly with only one process (*see* **Note 14**).

When standard addition method is applied, automated quantitation is not available. Integrated results for each sample and spiked/fortified samples together with corresponding spiking levels must be imported in spreadsheet program (excel or similar) where unknown concentrations of metabolites in samples can be calculated as *intercept/slope*.

4   Notes

1. This will help to avoid the growth of bacteria.

2. Classification of compounds in the current report is based on their reference concentration in serum according to the literature and HMDB database.

3. Compounds are present in biological fluids at different concentration ranges so their grouping is important in terms of quantitation.

4. In certain cases addition of minor amounts of base NaOH, KOH (for riboflavin, uric acid, xanthine, threonine, aspartate), or HCl acid (for 2-hydroxyisobutyric, inositol, 3-methylhistidine, cystine, xylose, tyrosine) and heating (for thymine, uracil) and/or sonication (for cysteine) is needed to assist dissolution.

5. Due to high concentration of buffer, addition of MeCN turns the solvent cloudy. Ultrasonic at room temperature assists in dissolvation.

6. Initial column pressure should not exceed 5300 psi; otherwise, system may be overpressure when eluent B reach 85%.

7. The choice of extraction solvent is based on average solvent content in MeCN, MeOH, and H_2O of standard mixtures.

8. Since there is no available analyte free matrix, the standard addition approach is performed in order to avoid limitations such as matrix effect, in cases were absolute concentration is required.

9. External calibration curve can be applied only when comparison between samples of the same or different batches and not absolute quantitation is required.

10. For the external calibration standards, standard 9 should be diluted 1:1 with 95:5 MeCN:H_2O and then proceed to serial dilutions.

11. Optimum column pressure ranges during the gradient: minimum 5000–5300 psi, maximum 11,000–12,500 psi. In case of higher pressures at the top end of the ranges clean the column as recommended.

12. According to our group tests that have been performed for system equilibration, four injections are the minimum for retention time and signal stabilization for serum.

13. According to tests that have been performed for system performance, 100 serum samples were considered as an upper limit in order to avoid variations due to column and cone contaminations from the matrix.

14. Always use mean calibration curve (before and after-batch calibration curve) for quantification. In cases when TargetLynx is used, just select standard samples of both curves together with real samples, QCs, and standard mixtures, and perform integration and quantitation by a single process.

References

1. Nicholson JK, Lindon JC, Holmes E (1999) 'Metabonomics': understanding the metabolic responses of living systems to pathophysiological stimuli via multivariate statistical analysis of biological NMR spectroscopic data. Xenobiotica 29:1181–1189

2. Nicholson JK, Lindon JC (2008) Systems biology: Metabonomics. Nature 455:1054–1056

3. Nicholson JK, Connelly J, Lindon JC et al (2002) Metabonomics: a platform for studying drug toxicity and gene function. Nat Rev Drug Discov 1:153–161

4. Theodoridis GA, Gika HG, Wilson ID (2011) Mass spectrometry-based holistic analytical approaches for metabolite profiling in systems biology studies. Mass Spectrom Rev 30:884–906

5. Zhou B, Xiao JF, Tuli L, Ressom HW (2012) LC-MS-based metabolomics. Mol BioSyst 8:470–481

6. Fiehn O, Robertson D, Griffin J et al (2007) The metabolomics standards initiative (MSI). Metabolomics 3:175–178

7. Theodoridis GA, Gika HG, Want EJ et al (2012) Liquid chromatography-mass spectrometry based global metabolite profiling: a review. Anal Chim Acta 711:7–16

8. Verpoorte R, Choi YH, Kim HK (2010) Metabolomics: will it stay? Phytochem Anal 21:2–3

9. Griffiths WJ, Koal T, Wang Y et al (2010) Targeted metabolomics for biomarker discovery. Angew Chem Int Ed Engl 49:5426–5445

10. Michopoulos F, Whalley N, Theodoridis G et al (2014) Targeted profiling of polar intracellular metabolites using ion-pair-high performance liquid chromatography and -ultra high performance liquid chromatography coupled to tandem mass spectrom. : applications to serum, urine and tissue extracts. J Chromatogr A 1349:60–68

11. Roberts LD, Souza AL, Gerszten RE et al Targeted metabolomics. Curr Protoc Mol Biol 98:302.1–302.24

12. Gika HG, Theodoridis GA, Vrhovsek U et al (2012) Quantitative profiling of polar primary metabolites using hydrophilic interaction ultrahigh performance liquid chromatography-tandem mass spectrometry. J Chromatogr A 1259:121–127

13. Schiesel S, Lämmerhofer M, Lindner W (2010) Multitarget quantitative metabolic profiling of hydrophilic metabolites in fermentation broths of β-lactam antibiotics production by HILIC-ESI-MS/MS. Anal Bioanal Chem 396:1655–1679

14. Coulier L, Bas R, Jespersen S et al (2006) Simultaneous quantitative analysis of metabolites using ion-pair liquid chromatography-electrospray ionization mass spectrometry. Anal Chem 78:6573–6582

15. Buescher JM, Moco S, Sauer U et al (2010) Ultrahigh performance liquid chromatography–tandem mass spectrometry method for fast and robust quantification of anionic and aromatic metabolites. Anal Chem 82:4403–4412

16. Virgiliou C, Sampsonidis I, Gika HG et al (2015) Development and validation of a HILIC-MS/MS multitargeted method for metabolomics applications. Electrophoresis 36:2215–2225

17. Sampsonidis I, Witting M, Koch W et al (2015) Computational analysis and ratiometric comparison approaches aimed to assist column selection in hydrophilic interaction liquid chromatography–tandem mass spectrometry targeted metabolomics. J Chromatogr A 1406:145–155

Ion Pair Chromatography for Endogenous Metabolites LC-MS Analysis in Tissue Samples Following Targeted Acquisition

Filippos Michopoulos

Abstract

A protocol for the preparation of tissue extracts for the targeted analysis of ca. 150 polar metabolites, including those involved in central carbon metabolism is described, using a reversed-phase ion pair U(H) PLC-MS method. Data collection enabled by multiple-reaction monitoring provides highly specific, sensitive acquisition of metabolic intermediates with a wide range of physicochemical properties and pathway coverage. Technical aspects are discussed for method transfer along with the basic principles of sample sequence setup, data analysis, and validation. General comments are given to help the assessment of data quality and system performance.

Key words Metabonomics, Targeted analysis, Mass spectrometry, Metabolites, Ion pair chromatography

1 Introduction

Late 1980s investigations that would lead to the formation of a new omic technology, subsequently called metabonomics, was established predominately based on ^{1}H NMR spectroscopy [1, 2]. The unparalleled reproducibility and quantitative nature of NMR spectroscopy, combined with its inherently high structural information content in combination with pattern recognition informatic tools opened up new horizons into understanding biological phenomena. Despite the relative lack of sensitivity that limits the use of the NMR technology to relatively highly abundant metabolites, the technique remained the workhorse of the fast-growing metabonomics field until early in the twenty-first century [3–7]. Advances in analytical instrumentation and particularly at the interface between mass spectrometry with separation techniques (liquid or gas chromatography, capillary electrophoresis, etc.) raised scientific interest in these hyphenated technologies as they

Georgios A. Theodoridis et al. (eds.), *Metabolic Profiling: Methods and Protocols*, Methods in Molecular Biology, vol. 1738, https://doi.org/10.1007/978-1-4939-7643-0_6, © Springer Science+Business Media, LLC, part of Springer Nature 2018

offered increased coverage of the metabolome and generally much greater sensitivity than ^{1}H NMR spectroscopy [8–12]. The opportunities offered by MS in combination with the plethora of available chromatographic substrates and formats open a new horizon in metabolite detection producing information-rich data that required the development of new bioinformatic tools to deconvolute the resulting complex data sets and visualize metabolic phenotypes [13–15]. Hypothesis-free, untargeted, metabolic profiling approaches greatly benefited from these technological advances in these hyphenated analytical tools which enabled the establishment of differential metabolic phenotypes across many research areas. However, successful deployment of LC-MS-based tools had to overcome challenges such as structure elucidation, metabolite ID determination, standardization, and reproducibility in untargeted metabonomics.

In the last decade, an increasing number of publications have reported the development of targeted methodologies for metabolite detection [16–19]. These assays utilize tandem mass spectrometers to selectively detect hundreds of known metabolic intermediates and are reproducible, amenable to standardization without requiring the often time-consuming structure elucidation required by untargeted methodologies. The ability to customize the selection of the metabolites to target specific biological process and pathways (e.g., central carbon metabolism) has enabled analysts to better optimize chromatographic parameters and improve the detection of highly polar metabolites such phosphorylated compounds and carboxylic acids. The depth of information provided from such "tailor-made" approaches has been driven by the need to address more mechanistic biological questions. In the case of targeted approaches for polar hydrophilic analytes, reversed-phase chromatography is poorly suited to their analysis, and chromatographic strategies employing hydrophobic interaction liquid chromatography (HILIC) [19–21] and ion-pair chromatography using [16–18] and to lesser extend derivatization protocols [22, 23] have been widely utilized.

Here we describe a targeted protocol using ion-pair chromatographic separation to resolve ~150 metabolic intermediates on a C18 column. The protocol has been optimized to provide semiquantitative data on tissue specimens although it has also been used to analyze urine, plasma, serum, and samples obtained from cell-based in vitro studies. The proposed methodology can also be adapted to address the demands for absolute quantification of analytes of particular interest, but for such a purpose, we advise the development of a separate assay to quantify just these metabolites.

2 Materials

Water of chromatographic grade purity (18.2 MΩ), organic solvents (acetonitrile, methanol, isopropanol), and mobile phase additives (acetic acid, tributylamine) of HPLC grade or higher should be used. The same quality criteria must be applied to the analytical standards to obtain the highest available purity.

2.1 Preparation of Analytical Standards, Test Mixture, and Infusion Solution

For all the analytical standards listed in Table 1, the appropriate amount is weighed in a 1.5 mL Eppendorf tube to result in a 1 mL 50 mM solution in MeOH/H$_2$O 50/50 v/v (stock solution). For amino acids and nucleotides, solubility is improved by the addition of small amounts of HCl, while other classes of metabolite may require small amounts of 1 M NaOH. Stock solutions must be diluted 1/100 v/v with HPLC grade water to produce a solution (dilution A) with nominal concentration of 500 μM which will be further diluted 1/50 with either HPLC water (dilution B) or MeOH/H$_2$O 50/50 v/v (infusion solution) to result final standard concentration of 10 μM. Stock solutions, dilutions A and B should be transferred to 1.8 mL cryovials and stored at −20 °C as reference material for future use.

An aliquot of 30 μL of the appropriate stock solution standards (*see* Table 1) can be mixed in a 15 mL Falcon tube and concentration adjusted to 100μM with HPLC water before dividing into 10 μL aliquots (test mixture 1–6) and storage at −20 °C for use for batch validation.

2.2 Solvent Preparation for Liquid Chromatography

2.2.1 Mobile Phase A

In a 15 mL Falcon tube, add 0.960 mL of acetic acid, 2.360 mL tributylamine, and 1.0 mL of MeOH. After a quick manual shake, the content of the tube is added to 990 mL of HPLC water in a 1 L solvent bottle, and then the falcon tube is rinsed with 5.68 mL of HPLC grade water before it is added to the mobile phase A bottle. Shake mobile phase A bottle manually for 1 min to ensure good solvent homogenization before connecting to the chromatographic system.

2.2.2 Mobile Phase B

In a 1 L solvent bottle mix 800 mL of MeOH with 200 mL of isopropanol and manually shake for 1 min before connecting to chromatographic system.

2.2.3 Syringe and Needle Wash

In a 0.5 L solvent bottle, mix equal volumes of isopropanol with acetonitrile and shake to ensure mixing.

2.2.4 Seal Wash Solvent

In a 0.5 L solvent bottle, mix 400 mL of HPLC water with 100 mL of isopropanol and shake to ensure mixing (*see* **Notes 1–3**).

2.3 Instrumentation

Sample analysis accordingly is performed on Thermo Ultimate 3000 RS pump combined with an Ultimate 3000 autosampler operating at 4 °C or instrumentation of similar specifications using

Table 1

List of metabolites measured with the current methodology, mass spectrometer parameters, and text mixtures composition

Metabolite name	Q1 mass (Da)	Q3 mass (Da)	RT (min)	DP (V)	EP (V)	CE (V)	CXP (V)	Test mixture
Pyruvic acid	87	43	6.4	−30	−10	−12	−1	1
Lactic acid	89	43	4.7	−40	−10	−19	−1	1
Glyoxylic acid	91	73	11.1	−18	−10	−13	−5	5
Malonic acid	103	59	9.7	−27	−10	−14	−5	3
Serine	104	74	0.8	−40	−10	−18	−5	1
Cytosine	110	67	0.8	−30	−10	−19	−5	4
Uracil	111	42	0.9	−26	−10	−30	−5	6
Creatinine	112	68	0.8	−55	−10	−25	−4	2
Proline	114	68	0.8	−55	−10	−18	−4	2
Maleic acid	115	71	9.9	−35	−10	−10	−3	3
Fumaric acid	115	71	11.3	−31	−10	−14	−5	4
Valine[1]	116	70	0.8	−70	−10	−15	−5	4
Succinic acid	117	73	9.9	−45	−10	−16	−5	1
Threonine	118	74	0.8	−45	−10	−15	−5	2
Benzoic acid	121	77	11.9	−30	−10	−19	−5	4
Nicotinic acid	122	78	9.5	−35	−10	−19	−5	3
Thymine	125.1	42	1.5	−35	−10	−22	−5	6
Pyroglutamic acid	128	84	5	−75	−10	−16	−5	
Mesaconic acid	129	85	11.3	−35	−10	−12	−5	2
Itaconic acid	129.1	85	10.6	−33	−10	−15	−5	6
Leucine	130	84	1.3	−55	−10	−18	−3	4
Isoleucine	130	84	1.2	−55	−10	−18	−3	2
Creatine	130	88	0.8	−35	−10	−14	−4	4
Asparagine	131.1	95	0.8	−31	−10	−19	−5	4
Glutaric acid	131.1	87	10.4	−35	−10	−19	−5	5
Aspartic acid	132	88	3	−40	−10	−19	−5	4
Malic acid	133	115	10.7	−30	−10	−16	−9	2
Adenine	134	107	1.2	−70	−10	−25	−5	2
Salicylic acid	137.1	65	13	−48	−10	−41	−5	6
Nitrophenol	138.1	108	10.4	−59	−10	−20	−5	6

(continued)

Table 1 (continued)

Metabolite name	Q1 mass (Da)	Q3 mass (Da)	RT (min)	DP (V)	EP (V)	CE (V)	CXP (V)	Test mixture
α-ketoglutaric acid	145	57	11.1	−30	−10	−18	−7	1
Adipic acid	145	83	10.8	−35	−10	−18	−5	3
α-ketoglutaric acid b	145	101	11.1	−35	−10	−12	−7	1
Glutamine	145.1	127.1	0.8	−45	−10	−16	−5	5
Glutamic acid	146	102	2.6	−40	−10	−19	−5	5
Hydroxy glutaric acid	147	85	10.7	−40	−10	−22	−5	4
Citramalic acid	147	87	10.6	−50	−10	−20	−5	6
Methionine	148	47	0.9	−35	−10	−25	−7	4
Guanine	150.1	133.1	1.6	−45	−10	−20	−5	5
Hydroxy phenyl acetic acid	151	107	9.9	−30	−10	−10	−5	2
Xanthine	151	108	1	−50	−10	−23	−10	
Histidine	154	93	0.8	−45	−10	−25	−4	4
Orotic acid	155	111	6.2	−30	−10	−18	−9	3
Pimelic acid	159	97	11.5	−40	−10	−20	−5	1
Pimelic acid 2	159	115	11.5	−40	−10	−20	−5	1
Indole-2-carboxylic acid	160.1	116.1	13.5	−40	−10	−20	−5	5
Mercapturic acid	162	84	9	−30	−10	−15	−10	
Coumaric acid	163.1	119.1	10.8	−45	−10	−21	−5	6
Phenylalanine	164.1	147.1	2.5	−48	−10	−19	−5	6
Phthalic acid	165	121	12.5	−40	−10	−16	−5	6
3-(2-Hydroxyphenyl)propionic acid	165.1	121.1	13	−44	−10	−21	−5	5
Methylxanthine	165.1	122.1	2	−47	−10	−27	−5	6
Quinolinic acid	166	122	11.9	−25	−10	−15	−5	
Phosphoenolpyruvic acid	167	79	11.8	−35	−10	−20	−5	1
Glyceraldehyde-3-phosphate	169	97	11.4	−22	−10	−16	−5	5
Dihydroxyacetone phosphate	169	97	13.7	−40	−10	−15	−5	
cis-Aconitic acid	173	85	11.9	−33	−10	−18	−5	3
Arginine	173.1	131.1	0.8	−40	−10	−19	−5	4
Shikimic acid	173.1	93	3.4	−40	−10	−25	−5	6
Citrulline	174.1	131.1	0.8	−30	−10	−21	−5	4

(continued)

Metabolite name	Q1 mass (Da)	Q3 mass (Da)	RT (min)	DP (V)	EP (V)	CE (V)	CXP (V)	Test mixture
Tyrosine	180	163	1.1	−45	−10	−20	−5	6
Sorbitol/mannitol	181	89	0.8	−58	−10	−19	−7	3
Pserine	184	97	6.7	−35	−10	−18	−5	1
Glycerate 3 phosphate	185	79	11.4	−35	−10	−44	−3	1
2-Trans-3-indolepyruvic acid	186.1	142.1	13.3	−30	−10	−29	−5	4
Kynurenic acid	188	144	11.8	−40	−10	−25	−9	
Citric acid	191	111	12	−30	−10	−18	−7	5
Citric acid specific	191	87	12	−35	−10	−26	−5	5
Isocitric acid specific	191	73	12.1	−35	−10	−32	−5	3
Glucuronic acid	193.1	113	3.7	−31	−10	−19	−5	5
Ferulic acid	193.1	134	11.1	−46	−10	−22	−5	6
Tryptophan	203.1	116	3.8	−50	−10	−25	−5	6
Xanthurenic acid	204	160	11.5	−45	−10	−20	−10	
Kynurenine	207	190	1.8	−40	−10	−12	−5	
p-Creatine	210	79	10.5	−35	−10	−35	−5	2
Pantothenic acid	218	88	9.6	−46	−10	−20	−5	3
N-Acetylglucosamine	220.2	119	0.8	−44	−10	−11	−5	6
Dinitrosalicylic acid	227	183	14.9	−50	−10	−22	−5	5
Ribose 5 P	229	97	4.7	−50	−10	−22	−5	1
Ribulose 5 P	229	97	5.8	−50	−10	−22	−5	2
Xylulose 5 P	229	97	5.8	−50	−10	−22	−5	5
Melatonin	231	216	11.2	−65	−10	−22	−5	
Cystine	239	120	0.8	−35	−10	−17	−5	2
Uridine	243	110	1.1	−60	−10	−25	−5	3
Palmitic acid	255.4	237.4	17.8	−80	−10	−29	−5	
Glucosamine-6-P	258.1	97	0.8	−35	−10	−26	−5	5
Glucose-6-P	259	97	4.3	−55	−10	−27	−5	1
Mannose-6-P	259	79	4.4	−55	−10	−70	−5	2
Galactose-1-P	259	241	4.7	−55	−10	−20	−5	4
Fructose-6-P	259	169	4.9	−55	−10	−17	−5	6
Glucose-1- P	259	241	5.3	−55	−10	−20	−5	3
Fructose-1-P	259	97	6.1	−55	−10	−65	−5	5

(continued)

Table 1 (continued)

Metabolite name	Q1 mass (Da)	Q3 mass (Da)	RT (min)	DP (V)	EP (V)	CE (V)	CXP (V)	Test mixture
Glycerate 1,3 bisphosphate	265	167	13.3	−40	−10	−20	−5	1
Adenosine	266	134.3	3.15	−38	−10	−18	−12	
Inosine	267	135	1.5	−80	−10	−37	−10	3
6-Phosphogluconic acid	275	79	11.3	−60	−10	−66	−5	1
Guanosine	282	150	1.6	−80	−10	−25	−5	3
Ophthalmic acid	288	195	6.4	−70	−10	−25	−5	
Arginosuccinate	289	132	2.75	−50	−10	−26	−12	
2′-Deoxycytidine monophosphate	306.1	79	7.6	−45	−10	−70	−5	5
Glutathione red	306.3	143.1	5.9	−50	−10	−26	−5	5
2′-Deoxyuridine monophosphate	307	111	8.9	−55	−10	−35	−5	1
Thymidine monophosphate	321	195	9.6	−55	−10	−25	−5	4
Cytidine monophosphate	322	97	6.6	−50	−10	−39	−5	3
Uridine monophosphate	323	97	7.9	−52	−10	−33	−5	3
Adenosine-3′5′- cyclic monophosphate	328	134	10.3	−80	−10	−40	−5	1
2′-Deoxyadenosine monophosphate	330.1	79	10	−55	−10	−60	−5	2
Fructose 1,6 bisphosphate	339	97	11.8	−60	−10	−28	−5	1
Guanosine-3′5′-cyclic monophosphate	344	150	9.8	−67	−10	−39	−5	4
2′-Deoxyguanosine monophosphate	346	79	9.3	−70	−10	−62	−5	6
Adenosine monophosphate	346.2	79	9.6	−70	−10	−62	−5	1
Inosine monophosphate	347	79	8.7	−55	−10	−85	−5	3
Guanosine monophosphate	362	79	8.7	−50	−10	−70	−5	3
S-5-Adenosyl-L-cysteine	369	282	1.3	−60	−10	−18	−5	
S-5-Adenosyl-L-cysteine 1	369	134	1.3	−60	−10	−35	−5	
Riboflavin	375.1	255.1	9.9	−55	−10	−26	−5	5
S-5Adenosyl-L-homocysteine	383	188	2.1	−40	−10	−24	−5	
2′-Deoxycytidine diphosphate	386	79	11.4	−50	−10	−88	−5	2

(continued)

Table 1 (continued)

Metabolite name	Q1 mass (Da)	Q3 mass (Da)	RT (min)	DP (V)	EP (V)	CE (V)	CXP (V)	Test mixture
Thymidine diphosphate	401	79	11.9	−70	−10	−80	−5	3
Uridine diphosphate	403	159	11.5	−60	−10	−41	−5	3
2′-Deoxyadenosine diphosphate	410	79	12.1	−70	−10	−85	−5	4
Butyryl CoA	418	79	15.1	−50	−10	−80	−5	5
Acetoacetyl CoA	424.8	382.8	14.1	−35	−10	−16	−4	5
Isovaleryl CoA	424.9	134	15.6	−50	−10	−45	−5	2
Malonyl CoA	425.5	404	14.3	−28	−10	−12	−4	2
Adenosine diphosphate	426.1	79	11.9	−80	−10	−88	−3	2
Methyl malonyl CoA	432.5	410.5	14.4	−30	−10	−10	−5	4
Folic acid	440	311	11.9	−75	−10	−32	−7	3
Guanosine diphosphate	442	79	11.5	−66	−10	−87	−5	5
2′-Deoxycytidine triphosphate	466	79	13.2	−80	−10	−105	−5	4
2′-Deoxyadenosine diphosphate	467	159	13.4	−60	−10	−50	−9	5
Thymidine triphosphate	481	159	13.5	−70	−10	−48	−5	2
2′-Deoxyadenosine triphosphate	490	79	13.5	−75	−10	−105	−5	1
Adenosine triphosphate	506.1	79	13.4	−90	−10	−106	−1	4
Guanosine triphosphate	522.1	79	13.3	−70	−10	−105	−5	2
ADP ribose	558.1	346.1	11.6	−80	−10	−33	−5	4
UDP glucose	565	323	11	−80	−10	−33	−5	3
UDP glucuronic acid	579	403	13.1	−60	−10	−32	−5	3
Glutathione ox	611.6	306.3	10.4	−80	−10	−31	−5	6
Nicotinamide adenine dinucleotide	662.3	540.1	7.8	−50	−10	−22	−9	1
Nicotinamide adenine dinucleotide reduced	664.3	79	12	−110	−10	−120	−3	2
Nicotinamide adenine dinucleotide phosphate	742.2	620.1	11.7	−60	−10	−22	−11	5
Nicotinamide adenine dinucleotide phosphate reduced	744.3	79	13.4	−110	−10	−118	−3	1
Coenzyme A (CoA)	766	79	14	−150	−10	−130	−5	
Flavin-adenine-dinucleotide	784.5	97	13.3	−85	−10	−53	−5	6
Acetyl CoA	808.3	79	14.2	−85	−10	−53	−5	1

(continued)

Table 1 (continued)

Metabolite name	Q1 mass (Da)	Q3 mass (Da)	RT (min)	DP (V)	EP (V)	CE (V)	CXP (V)	Test mixture
Proprionyl CoA	822	79	14.6	−130	−10	−120	−5	6
Crotonyl CoA	834	79	14.4	−150	−10	−90	−5	4
Isobutyryl CoA	836.1	408	15.1	−155	−10	−55	−5	3
OH Butyryl CoA	852	79	14.1	−150	−10	−90	−5	6
Succinyl CoA	866.2	79	14.5	−130	−10	−120	−5	6
Hydroxymethyl glutaryl CoA	910.2	79	14.4	−150	−10	−120	−5	6

[1]Valine very weak ionization in negative ESI

Ribose-5-P, ribulose-5-P, xylulose-5-P, glucose-6-P, and so forth correspond to ribose-5-phosphate, ribulose-5-phosphate, xylulose-5-phosphate, glucose-6-phosphate

DP declustering potential, *EP* entrance potential, *CE* collision energy, *CXP* collision exit potential

a Acquity HSS T3 UPLC column (Waters Corp, 2.1 × 100 mm, 1.8 μm particle size) with column temperature maintained at 60 ± 0.5 °C. Spectrometric data is acquired on ABSCIEX 4000 triple quadrupole instrument operation in negative electrospray ionization acquisition mode.

3   Methods

3.1   U(H)PLC

Analysis is performed using a 5 μL injection of sample with gradient elution at a flow rate of 400 μL/min with a binary solvent mixing schedule of Mobile phase B, over Mobile phase A: 0 min, 0% B; 0.5 min, 0% B; 4 min, 5% B; 6 min, 5% B; 6.5 min, 20% B; 8.5 min, 20% B; 14 min, 55% B; 15 min, 100% B; 17 min, 100% B; 18 min, 0% B; and 21 min 0% B.

3.2   Mass Spectrometry

To obtain the optimum detection parameters, each metabolite (infusion solution) is infused via a syringe pump (Harvard Apparatus 22) at flow rate of 10 μL/min, and the optimized parameters for declustering potential, entrance potential, collision energy, and collision exit potential are given in Table 1. Interface source parameters should be optimized at the flow rate (0.4 mL/min) of the intended chromatographic separation. For this purpose 10 μL/min of each metabolite (dilution A) should be infused via a syringe pump to a T-connector where it should be combined with 0.39 mL/min 90% mobile phase A via the UHPLC system delivering to the ESI source a final metabolite concentration of 12.5μM. The optimal ESI source parameters should be as follows: ion spray voltage −4.0 kV, temperature 550 °C, collision gas 5, curtain gas

30, ion source gas1 60, and ion source gas2 50. Gas values are arbitrary units.

3.3 Tissue Extraction and Preparation

Tissue samples must be processed and extracted from frozen to reduce endogenous metabolite degradation. For the present protocol, we can use a combined extraction homogenization approach that is performed using a Precellys 24 system with an attached temperature control unit. For soft tissue samples, we advise the use of CK14 tubes, while for harder tissue, the CK28R format is more appropriate. Extraction and homogenization is achieved in the appropriate tube format from a 50 mg of frozen tissue with 1 mL ACN/MeOH/H_2O 40/40/20 v/v/v. Tubes must be shaken at 5000 rpm for 20 s, and the process must be repeated three times with an intermittent pause of 30 s between each repeat. While the extraction and homogenization is taking place, the extraction chamber unit must be at the lowest possible temperature and not more than 10 °C. A clear supernatant is obtained after centrifugation at 10691 × g for 5 min at 0 °C, and this is transferred to cryovial for storage at a minimum −20 °C until analysis. The extraction-homogenization procedure is repeated with fresh solvent as described above, and the resulted clear supernatant is combined in the same cryovial.

3.4 Pre-Analytical Considerations

Prior to LC-MS analysis, a minimum of a 1 in 10 dilutions with HPLC water must be performed to avoid excess carryover of highly abundant metabolites across injections. For different tissues, an exploratory dilution experiment can be employed using different dilution extract/total sample volume ratios (1/10, 1/20 v/v, etc.) to define the optimal dilution for a given study. Samples after water dilution must be quickly vortexed and centrifuged at 2250 × g for 10 min at 4 °C.

For batch validation and metabolite confirmation, a pooled sample (QC) is prepared by mixing equal aliquots of the individual extracts, and this is treated the same way as the study sample extracts. The QC sample is also used to condition the analytical platform at the beginning of the sample sequence as well as being injected at regular intervals during the analytical sequence to assess metabolite analytical reproducibility.

Metabolite identification and retention time confirmation is obtained by cross comparison between text mixture, spiked QC, and individual samples. The test mixture is an aqueous sample consisting of test mixtures 1–6 diluted 1/20 v/v with HPLC water. The spiked QC sample is a test mixture sample prepared in the matrix (QC) of interest.

3.5 Sample Sequence

The sample sequence should be set up appropriately to provide a clear measure of the data quality and system performance. Gradient blank, solvent blank, test mixture, conditioning samples, quality controls (QC), and spiked QC are additional to individual sample injections that help in assessing the analysis quality. The use of these injections and a brief description of the sample sequence are given here:

1. A gradient blank at the beginning of the sample sequence helps to identify any signals/impurities/contaminants related to the solvents used for chromatography or prolonged carry-over in the system. The gradient blank is an acquisition where no injection is performed and the chromatographic system runs the gradient solvent schedule (*see* **Notes 4** and **5**). As contaminants can be specific to different batches of solvent, this is an important pre-analysis check as poor quality solvents can adversely affect the analysis (*see* **Notes 1–3**).

2. The use of two solvent blank injections after the gradient blank helps the identification of signals related to solvent in which the samples have been reconstituted/diluted. Consecutive injections of solvent blank injections enable the assessment of the severity and persistence of any carry-over.

3. System conditioning, which is essential to stabilize analyte retention times, is performed with the same matrix as that of the analysis sample set. The pooled sample, described in the tissue extraction and preparation section above, is recommended to be used unless volume limitations due to small sample size prohibit its use. In these circumstances, a matrix sample of the same origin from a different study can be substituted. System conditioning improves reproducibility of detected signals and the number of conditioning injections is matrix and instrument saturation related. A minimum of 7–10 conditioning injections is recommended (*see* **Note 6**).

4. Test mixture and spiked QC injections are the first steps of batch quality validation. After system conditioning, two injections of the test mixture followed by a spiked QC injection enables retention time confirmation and assessment of the overall system performance by examination of the resulting metabolite resolution in the ion chromatogram. The first text mixture injection does not always provide the optimal chromatographic separation so two to three consecutive injections are recommended.

5. Due to the relatively high metabolite concentrations (5 μM for each) in both test mixture and spiked QC, which may result in some modest carry-over on the following injection, a further two to three system conditioning injections (as per point 3 above) are recommended to re-equilibrate the chromatographic system for analysis of the sample set under investigation.

6. QC injections are the foundation of the statistical analysis and the assessment of the analytical reproducibility for each metabolite detected. The sample analysis should start and finish with a

QC injection to enable quality assessment at the beginning and end of the individual sample analysis. A minimal of five QC injections are required to obtain sufficient robustness for the statistical analysis. Therefore QC injections are performed at regular intervals, interspersed between the individual test samples (*see* **Note 7**).

7. Samples are analyzed in random order to reduce data bias, in blocks of five to ten, sandwiched between two consecutive QC injections. The number of injections in a sample set defines the length of the sample block required to achieve the minimum of five QC injections needed for robust statistical results.

8. Upon completion of the last QC injection, a spiked QC and a test mixture injection should be obtained to assess any metabolite retention time drift and intensity loss during the analysis along the sample sequence.

9. Finishing the sample sequence with two injections of the solvent blank and a blank gradient helps to explore carry-over across the analytical batch as well as any increase or decrease on the background signal (threshold) essential for peak integration.

3.6 **Data Analysis**

Raw spectrometric data are processed with MultiQuant software to obtain peak areas for each of the detected metabolites across the sample set. The first step prior to peak integration is the visual confirmation of the retention time for each metabolite peak by comparing the trace obtained for the test mixture, spiked QC, and the individual samples. Given the inability to obtain biological matrix free from the endogenous metabolites being determined, the solvent blank injection data can be used to define the background threshold values for each metabolite to ensure signals are associated with actual metabolite presence and not the background. Smoothing factor and peak splitting are two additional parameters that must be customized using the sample injections in order to obtain accurate and reproducible peak integration. The peak integration report can be exported as a text file from the MultiQuant software and further processed in customer optimized visualization and statistical software packages. For example, in Excel univariate statistical analysis (f-test, t-test, ANOVA) can be performed to validated significant differences across meaningful group comparison on metabolite level very simply. It is also essential to extend the univariate analysis and combine with analytical reproducibility data (coefficient of variation values, CV) to reinforce statistical significance (*see* **Note 8**). Data normalization is also essential to reduce bias and trends due to analysis order or loss of signal intensity since no internal standards are used in this analysis protocol. Median value normalization is highly regarded as an adequate approach for this type of data set. Typical validation criteria for detecting

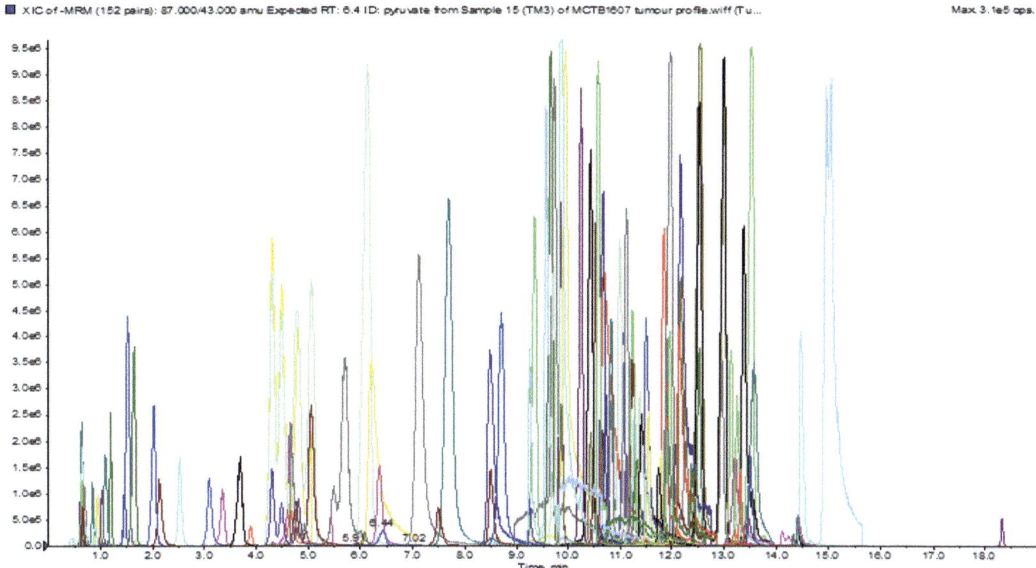

Fig. 1 Typical ion chromatogram of a text mixture injection obtained following the proposed chromatographic separation

important differences in metabolite concentrations across two groups of samples can be regarded as the complete accomplishment of the following three: (1) absolute natural logarithmic fold change >0.5, coefficient of variation <30%, and p-value < 0.05. Figure 1 provides a representative ion chromatogram trace of a text mixture injection following the gradient elution sequence of the proposed protocol.

4   Notes

1. Prepare fresh mobile phase A before every analytical batch of samples. This helps in reducing background signals as well as avoiding bacterial growth in the aqueous LC solvent.

2. Replace mobile phase B every month to avoid retention time shift due to evaporation-induced changes in solvent composition.

3. Monitor pressure buildup across the analytical batch to avoid deterioration of chromatographic performance.

4. Always prime solvent lines before commencing a new batch analysis. This helps to reduce the formation of air bubbles in the LC system.

5. Ensure enough needle and syringe wash is available to reduce carry-over between sample injections.

6. Clean the ESI source after prolonged periods of use to restore mass spectrometer sensitivity.

7. Data quality assessment is essential prior to data analysis steps. Visual cross-comparison of ion chromatograms (overlay them) of two test mix injections: one acquired at the beginning and the second one at the end of the sample sequence must be utilized to assess signal intensity loos or retention time drift and reproducibility of the chromatographic resolution. Similar examination must be performed using the first and last QC sample injections. If signal intensity drops by more than 30%, it is advised not to proceed further the data analysis. Similarly, measures must be taken if retention time drifts more than 20 s across the analysis time. Signal loss or retention time drift must not be assessed using conditioning injection data.

8. Data visualization with multivariate statistical tools such as principal component analysis (PCA) can be utilized to more comprehensively assess data quality by examining QC group injections forming a tight cluster on the scores plot.

References

1. Bales JR et al (1998) Metabolic profiling of body fluids by proton NMR: self-poisoning episodes with paracetamol (acetaminophen). Magn Reson Med 6(3):300–306

2. Nicholson JK, Wilson ID (1989) High resolution proton magnetic resonance spectroscopy of biological fluids. Prog Nucl Magn Reson Spectrosc 21(4–5):449–501

3. Barzilai A et al (1991) Phosphate metabolites and steroid hormone receptors of benign and malignant breast tumors. A nuclear magnetic resonance study. Cancer 67(11):2919–2925

4. Gavaghan CL et al (2000) An NMR-based metabonomic approach to investigate the biochemical consequences of genetic strain differences: application to the C57BL10J and Alpk:ApfCD mouse. FEBS Lett 484(3):169–174

5. Kurhanewicz J et al (1995) Citrate as an in vivo marker to discriminate prostate cancer from benign prostatic hyperplasia and normal prostate peripheral zone: detection via localized proton spectroscopy. Urology 45(3):459–466

6. Lynch MJ, Nicholson JK (1997) Proton MRS of human prostatic fluid: correlations between citrate, spermine, and myo-inositol levels and changes with disease. The Prostat 30(4):248–255

7. Marx A et al (1996) Determination of the fluxes in the central metabolism of Corynebacterium glutamicum by nuclear magnetic resonance spectroscopy combined with metabolite balancing. Biotechnol Bioeng 49(2):111–129

8. Gika HG et al (2010) Does the mass spectrometer define the marker? A comparison of global metabolite profiling data generated simultaneously via UPLC-MS on two different mass spectrometers. Anal Chem 82(19):8226–8234

9. Dunn WB, Ellis DI (2005) Metabolomics: current analytical platforms and methodologies. TrAC Trends Anal Chem 24(4):285–294

10. Lenz EM, Wilson ID (2007) Analytical strategies in Metabonomics. J Proteome Res 6(2):443–458

11. Lindon JC, Nicholson JK (2008) Analytical technologies for metabonomics and metabolomics, and multi-omic information recovery. TrAC Trends Anal Chem 27(3):194–204

12. Theodoridis G, Gika HG, Wilson ID (2008) LC-MS-based methodology for global metabolite profiling in metabonomics/metabolomics. TrAC Trends Anal Chem 27(3):251–260

13. Shulaev V (2006) Metabolomics technology and bioinformatics. Brief Bioinform 7(2):128–139

14. Smith CA et al (2006) XCMS: processing mass spectrometry data for metabolite profiling using nonlinear peak alignment, matching, and identification. Anal Chem 78(3):779–787

15. Xia J et al (2012) MetaboAnalyst 2.0—a comprehensive server for metabolomic data analysis. Nucleic Acids Res 40(W1):W127–W133

16. Lu W, Bennett BD, Rabinowitz JD (2008) Analytical strategies for LC–MS-based targeted metabolomics. J Chromatogr B 871(2):236–242

17. Buescher JM et al (2010) Ultrahigh performance liquid chromatography– tandem mass spectrometry method for fast and robust quantification of anionic and aromatic metabolites. Anal Chem 82(11):4403–4412

18. Michopoulos F et al (2014) Targeted profiling of polar intracellular metabolites using ion-pair-high performance liquid chromatography and-ultra high performance liquid chromatography coupled to tandem mass spectrometry: applications to serum, urine and tissue extracts. J Chromatogr A 1349:60–68

19. Gika HG et al (2012) Quantitative profiling of polar primary metabolites using hydrophilic interaction ultrahigh performance liquid chromatography–tandem mass spectrometry. J Chromatogr A 1259:121–127

20. Schiesel S, Lämmerhofer M, Lindner W (2010) Multitarget quantitative metabolic profiling of hydrophilic metabolites in fermentation broths of β-lactam antibiotics production by HILIC–ESI–MS/MS. Anal and Bioanal Chem 396(5):1655–1679

21. Yuan M et al (2012) A positive/negative ion-switching, targeted mass spectrometry-based metabolomics platform for bodily fluids, cells, and fresh and fixed tissue. Nat Protoc 7(5):872–881

22. Kloos D et al (2014) Analysis of biologically-active, endogenous carboxylic acids based on chromatography-mass spectrometry. TrAC Trends Anal Chem 61:17–28

23. Kloos D et al (2012) Derivatization of the tricarboxylic acid cycle intermediates and analysis by online solid-phase extraction-liquid chromatography–mass spectrometry with positive-ion electrospray ionization. J Chromatogr A 1232:19–26

Chapter 7

LC-MS Untargeted Analysis

Elizabeth J. Want

Abstract

LC-MS untargeted analysis is a valuable tool in the field of metabolic profiling (metabonomics/metabolomics), and the applications of this technology have grown rapidly over the past decade. LC-MS offers advantages over other analytical platforms such as speed, sensitivity, relative ease of sample preparation, and large dynamic range. As with any analytical approach, there are still drawbacks and challenges to overcome, but advances are constantly being made regarding both column chemistries and instrumentation. There are numerous untargeted LC-MS approaches which can be used in this ever-growing research field; these can be optimized depending on sample type and the nature of the study or biological question. Some of the main LC-MS approaches for the untargeted analysis of biological samples will be described in detail in the following protocol.

Key words LC-MS, Untargeted, Mass spectrometry, Liquid chromatography, Metabolic profiling

1 Introduction

Metabolic profiling involves the measurement of low molecular weight metabolites, typically <1 kDa in a biological sample, such as a biofluid (e.g., urine, serum) or tissue. These measurements can offer valuable insights into responses to therapeutic interventions, disease diagnosis and progression, as well as the effects of ageing, diet, and exercise on an individual [1–3].

The complexity of biological samples means that no single analytical approach will provide complete metabolome coverage. Typically, a combination of NMR spectroscopy, GC-MS, and LC-MS will be employed to enrich the metabolome information recovered [4–6]. Improvements in instrument sensitivity and resolution, as well as ever-growing databases for the structural identification of metabolites, mean that greater metabolome coverage can now be achieved [7].

LC-MS is a sensitive tool for metabolic profiling. Separation of metabolites in a sample prior to mass spectrometric (MS) analysis reduces the complexity of the sample and hence the risk of ion suppression. Chromatographic separation can also aid in discriminating

Georgios A. Theodoridis et al. (eds.), *Metabolic Profiling: Methods and Protocols*, Methods in Molecular Biology, vol. 1738, https://doi.org/10.1007/978-1-4939-7643-0_7, © Springer Science+Business Media, LLC, part of Springer Nature 2018

between many isobaric species, which, when combined with high accuracy MS measurements, can improve considerably the structural elucidation of potential biomarkers. The breadth of LC column chemistries and types of mass spectrometers available mean that LC-MS is a versatile technique for both untargeted metabolic profiling and targeted metabolite analyses. Further, the advent of ultra(high) performance liquid chromatography (UPLC or UHPLC) has significant improved chromatographic resolution and hence metabolite coverage [8–11]. Here, we will focus on the application of LC-MS for the untargeted analysis of biofluids and tissues.

Whereas targeted metabolite analysis focuses on a particular class of molecules or pathway, offering sensitive, accurate, and quantitative measurements [12, 13], untargeted LC-MS analyses can be invaluable for hypothesis generation and biomarker discovery [14]. The keys to these untargeted approaches are (a) unbiased sample preparation, i.e., not favoring the extraction of a particular type or class of molecule, and (b), where possible, unbiased LC-MS analysis, where metabolites spanning a range of classes and polarities can be separated and detected in one or a handful of chromatographic analyses.

Typically, untargeted LC-MS studies employ reversed-phase (RP) chromatographic techniques [15–19]. RP chromatography enables the separation and detection of a range of moderately polar to nonpolar metabolites. However, many biofluids or tissue samples contain a plethora of more polar molecules, which will not be retained and detected using RP approaches. Hydrophilic interaction liquid chromatography (HILIC) has grown in applications and value for studies including metabolic profiling [20, 21]. HILIC, akin to normal phase chromatography, enables the separation and detection of polar and nonpolar molecules, with greater retention of polar molecules than RP technologies. Thus, HILIC can be considered a complementary technique to RP chromatography and to this end is often used in parallel for metabolic profiling studies.

Mass spectrometers of choice for untargeted LC-MS studies are time-of-flight (ToF) or quadrupole-time-of-flight instruments (Q-ToF). Advantages of these mass spectrometers include high mass resolution and mass accuracy, sensitivity, fast scan speed, and large dynamic range. The ability of Q-ToF mass spectrometers to perform MS/MS experiments to aid in structural elucidation of potential biomarkers makes them invaluable in untargeted metabolic profiling.

Here, untargeted LC-MS protocols are described for the analysis of biological samples using both RP and HILIC chromatography. Extraction of metabolites from different biological samples is described. The analysis of tissue samples is covered in more detail in Chapter 17. The importance and implementation of data quality assessment, such as through the use of quality control samples, are introduced but covered in more detail in Chapter 2.

2 Materials

2.1 Samples

In metabolic profiling, biological samples are commonly analyzed, e.g., urine, serum/plasma, tissue, cell extracts, and cell media (*see* **Notes 1–3**).

2.2 Standards and Chemicals

1. System suitability/test mix.

   (a) It is advised to include a mixture of compounds, which can be termed "system suitability mix or test mix" (*see* Subheading 3.3) in order to assess chromatographic and mass spectrometric performance. This can be commercially obtained or designed in-house to encompass metabolite classes of interest.

2. Metabolite extraction solvents.

   (a) Plasma/serum: methanol, acetonitrile.

   (b) Urine (HILIC analysis): acetonitrile.

   (c) Tissue: methanol/dichloromethane/MTBE/water.

3. UPLC-MS mobile phases.

   (a) LC-MS grade water, acetonitrile, methanol, isopropanol—exact compositions depend on nature of analysis (*see* Subheading 3.3; LC gradients) (*see* **Note 10**).

   (b) Mobile phase additives: formic acid, ammonium acetate, ammonium formate.

4. Lockmass and calibration solutions: leucine-encephalin, sodium formate.

2.3 Common Equipment

1. Pipettes.
2. Pipette tips.
3. Eppendorf tubes.
4. Bead beater polypropylene tubes.
5. MS vials.
6. MS well plates.
7. Sealing cap mats.
8. Glass bottles.

2.4 LC-MS Systems

1. LC or UPLC chromatography system, e.g., Acquity Ultra Performance LC.
2. Analytical columns—specific to chromatographic method, e.g., C18, C8, HILIC, HSS.
3. Mass spectrometer, e.g., Q-ToF.

2.5 Other Instrumentations

1. Benchtop centrifuge.

2. Precellys bead beater or similar.

3. Vacuum evaporator.

4. Ultrasonic bath.

5. Vortex mixer.

2.6 Data Processing Software

1. Vendor-supplied software or freeware (e.g., XCMS, MZMine 2) for preprocessing of raw data. This includes peak detection, alignment, and normalization (Subheading 3.6).

2. R and associated software packages.

3. Excel or similar.

4. SIMCA, MATLAB, or similar for multivariate analysis.

5. PRISM or JMP or similar for univariate analysis.

3 Methods

LC-MS untargeted analysis can be divided into multiple key steps (Fig. 1) as follows: sample preparation, chromatographic separation, and mass spectrometric detection, followed by data analysis. Sample preparation to extract metabolites is dependent on sample type. The most common samples used in untargeted LC-MS analyses are biofluids, e.g., urine, serum, plasma, as well as tissue (liver, heart, brain, etc.), cell extracts, and cell media. Metabolites are extracted from these samples and the extracts analyzed by LC-MS or UPLC-MS, often using a Q-ToF mass spectrometer, ideally employing both positive and negative electrospray ionization. Key to sample preparation is the removal of particulates and proteins from the samples as these can affect chromatographic performance and may even result in column blockage. Sample preparation for untargeted LC-MS analysis should not favor specific types of molecules but rather should aim for a broad extraction of metabolite classes, unless there is a specific interest in a class of molecules. It may be that the preparation method extracts polar metabolites into an aqueous fraction and nonpolar metabolites into an organic fraction, which are then analyzed using separate chromatographic methods, as in the case of tissue samples. This has benefits of reducing sample complexity and may improve detection, quantification, and identification.

Important aspects of an untargeted LC-MS analysis are:

System Suitability or Test Samples. Typically, 10–15 compounds will be used in a system suitability mix, which will be run at the start of the analysis (*see* **step 8**, Subheading 3.3) to assess instrument performance (e.g., chromatographic peak shape and retention time, mass accuracy, and detector response) before the analysis of study samples. Parameters such as detector response are prone

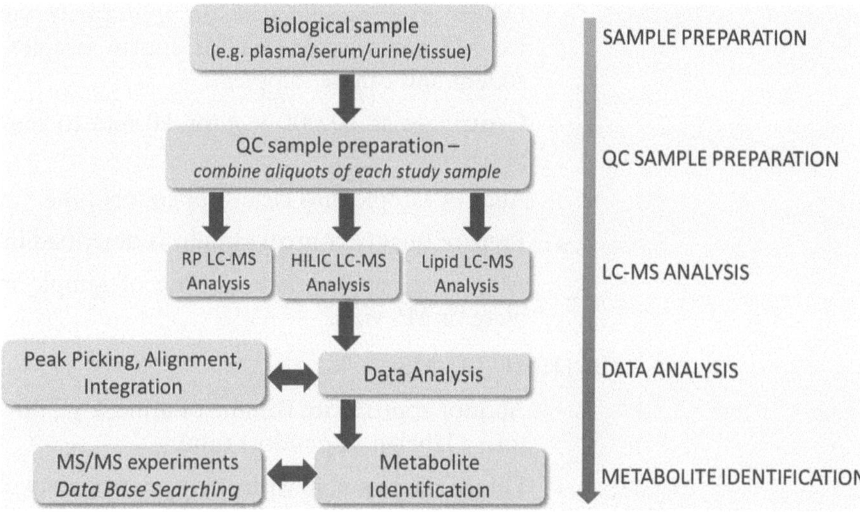

Fig. 1 LC-MS untargeted workflow showing the main steps of the process. These steps are described in this protocol

to change over the analytical run, and so this mix can also be run at the end of the sample analysis to assess these changes. For a more detailed description, the reader is directed to Chapter 2.

Quality Control Samples: Quality control (QC) samples are the key to successful, robust untargeted LC-MS analyses. These samples are representative of the study sample set and are usually made by mixing small aliquots (e.g., 10–50 μL) from all study samples. QC samples are used for (a) conditioning of the LC column and (b) assessment of data quality. Subheading 3.3 **step 9** describes the setup for QC samples in the run. For a more detailed description, the reader is directed to Chapter 2.

Sample Randomization: Study samples should be randomized within a batch to avoid bias which may arise from changes in the system over the course of the run, e.g., decrease in detector sensitivity and drifts in chromatographic retention times or mass accuracy. Samples can be randomized using in-house scripts, online software (https://www.randomizer.org/; https://www.random.org/lists/), or randomized block design in the case of larger studies. For a more detailed description, the reader is directed to Chapter 2.

3.1 Sample Preparation

1. Urine [22]

    (a) RP LC-MS:

        – Aliquot an appropriate volume of urine, e.g., 50–100 μL, into a labeled Eppendorf tube.

- Dilute urine sample with an appropriate volume of LC-MS grade water (1:1 for human samples, 1:3 for rodent and canine samples).
- Centrifuge at 10,000 × g for 10 min to remove any precipitate.
- Aliquot sample into clean Eppendorf tube.
- Prepare quality control sample as described in **step 5**.
- Aliquot an appropriate volume of sample into well plate or MS vials.

(b) HILIC LC-MS:

- Aliquot appropriate volume of urine, e.g., 50–100 μL, into a labeled Eppendorf tube.
- Dilute urine sample with appropriate volume of acetonitrile (1:1 for human samples, 1:3 for rodent and canine samples).
- Centrifuge at 10,000 × g for 10 min to remove precipitate.
- Aliquot sample into clean Eppendorf tube.
- Prepare quality control sample as described in **step 5**.
- Aliquot appropriate volume of sample into well plate or MS vials.

2. Serum/plasma

(a) The key to extracting metabolites from serum/plasma is the precipitation of proteins. Here, a simple method involving protein precipitation with methanol is suggested [23].

   Method 1: Preparation for RP LC-MS using methanol extraction

(b) Aliquot an appropriate volume of serum/plasma into a labeled Eppendorf tube.

(c) Add cold methanol in a ratio of 3:1, e.g., 150 μL methanol:50 μL plasma.

(d) Leave at −20 °C for at least one hour; it is acceptable to leave samples at this temperature overnight.

(e) Centrifuge at 10,000 × g for 10 min.

(f) Remove supernatant.

(g) Dry in vacuum concentrator.

(h) Resuspend in water or RP starting mobile phase (*see* **Note 6**).

3. Tissue sample preparation for RP and HILIC analysis [18, 24, 25]

(a) Weigh 50 mg (+/− 5 mg) tissue section and place in a) 2 mL polypropylene bead-beater tubes if using a Precellys bead beater (or similar) or b) 2 mL Eppendorf tubes if using a Qiagen tissue lyser (or similar). Keep tubes on ice before, during, and after the extraction process.

(b) Homogenize the sample with 1.5 mL prechilled methanol/water (1:1) using either a Precellys bead beater or Qiagen tissue lyser following the instructions below. The number of bead beating cycles will vary depending on the tissue type (*see* **Notes 4–5**).

Method	Requirements	Instruction
Precellys bead beater	Add 1 mm zirconium beads—100 μL in tube	40 s per cycle, cool on ice; 40 s for subsequent cycles
Qiagen tissue lyser	Stainless steel beads	25 Hz speed, 5 min cycle

(c) Centrifuge the sample at 4 °C for 10 min at 10,000 × *g*.

(d) *To prepare the aqueous extracts,* aliquot the supernatant into a clean Eppendorf tube. Retain the pellet in the bead-beater tube on ice for subsequent organic extraction (*see* **step 9**) (*see* **Note 7**).

(e) Take separate 250 μL aliquots for RP and HILIC LC-MS analyses.

(f) Take 150 μL of each sample to form the QC sample.

(g) Dry aqueous extracts in a Savant vacuum concentrator for ~180 min at 45 °C using the V-AQ (vacuum aqueous) mode.

(h) The sample can then be treated in one of the following ways:
   – Resuspension in starting mobile phase or similar and immediate analysis.
   – Storage at −40 °C or lower until analysis.

(i) *To prepare* the *organic extracts,* add 1.5 mL prechilled solution of dichloromethane/methanol (1:3) to the pellet from **step 4**.

(j) Homogenize the sample using either a Precellys bead beater or Qiagen tissue lyser, following the instructions in **step 2** (*see* **Notes 8** and **9**).

(k) Take separate 250 μL aliquots for RP (lipid) and HILIC LC-MS analyses.

(l) Take 150 μL of each sample to form the QC sample.

(m) Store at −40 °C or lower until analysis.

3.2 Preparation of LC-MS Calibration Solution

Prepare a fresh stock solution of sodium formate; 0.1 mg/mL in water. Add 1 mL of stock solution to 9 mL of isopropanol (IPA) to obtain a working solution of 0.01 mg/mL. This solution can be stored at 4 °C for a number of weeks. Note that an alternative calibration solution can be used instead, and many instrument vendors may recommend one.

3.3 LC-MS Analysis

1. Transfer samples to LC-MS vials or polypropylene 96-well plates. If samples have been stored in vials or well plates prior to analysis, centrifuge at 1350 × *g* for 5 min at 4 °C (ideally) or room temperature.

2. Place vials (see note 14) or well plates in the LC autosampler. The autosampler temperature (*see* **Notes 12** and **13**) should be maintained at 4 °C for all analyses except for lipid analysis, where the temperature should be 8–10 °C to avoid precipitation of the lipids. Samples should be stable in the autosampler for 24–48 h (see note 15), but the stability of individual metabolites over this time period will naturally vary. Stability of metabolites can be assessed in the different biological sample types, e.g., by using the QC samples.

3. Create the sample analysis list. Choose the LC gradient and MS settings as described in Subheadings 3.3 and 3.4, respectively. Column temperature will be dependent on the analysis and will vary from 40 °C for analysis of aqueous metabolites in urine samples to 55 °C for lipid analysis. Column temperature has been optimized based on the mobile phases employed for the LC gradient.

4. Select electrospray ionization mode.

5. Ensure that the instrument is set up for the optimum accurate mass by infusing a solution of a reference compound, such as leucine encephalin into the mass spectrometer. Follow the instrument-specific instructions for setup and acceptance criteria.

6. Calibrate the instrument prior to analysis using a solution of the appropriate concentration of sodium formate (Subheading 3.2) or similar calibration solution. Follow the instrument specific instructions for calibration and acceptance criteria. Calibration points should be spread as evenly as possible over the mass range that is being acquired for the analysis (typically 50–1000 or 1200 m/z for metabolic profiling). In general, the residual mDa for each individual calibration point should be <1.5 mDa, with the majority of calibration points having residuals of <0.5 mDa. With most instruments, this can be performed using a setup wizard or manually depending on the preferences of the instrument operator.

7. Start with a blank sample to assess the condition of the column. This blank sample can be water or a solvent mixture similar to the starting mobile phase conditions. If the LC column appears "dirty," i.e., shows a high background signal, then a sequence of blank samples is recommended. Background signal levels can be monitored until an acceptable level is reached.

8. Run the system suitability/test mix to assess peak intensity, retention times, and mass accuracy.

9. Inject quality control (QC) samples (*see* **Note 11**) in order to condition the column before study sample analysis. At least 10 QC samples are recommended for RP analysis of urine and plasma/serum. This may need to be increased to ~15 QC samples for RP analysis of tissue samples and for HILIC analysis in general. These QC samples can be assessed by visually overlaying and monitoring peak intensity, retention times, and mass accuracy. Conditioning QC samples can also be used to assess the appropriate injection volume needed in order to obtain peaks that are not overloaded and, in the same manner, sample dilution factors. An example of overlaid QC sample chromatograms is shown in Fig. 2.

10. In addition, set up untargeted MS/MS experiments, e.g., DDA experiments using at least one conditioning QC sample (*see* Subheading 3.5).

11. Randomize the study samples, add to the sample analysis list, and inject a QC sample every 5–10 samples. Aim for at least 10 QC samples per study batch in order to be able to perform appropriate statistics. End the sample batch with a QC sample, followed by a blank sample to assess carryover and a final system suitability/test mix injection to assess system changes over the run.

12. Injection volume will depend on sample type and instrumentation but usually ranges from 2 to 10 μL.

13. When satisfied with the performance of the system, using the system suitability mix and the conditioning QC samples, start the analysis of the study samples, including within run QC samples.

3.3.1 LC Gradients

1. Examples of UPLC gradients for urine, serum, and tissue samples by both RP and HILIC analysis are shown in the following tables: Tables 1a, 1b, 1c, 1d, 1e, 1f, and 1g. The column dimensions and details are also provided. Note that these gradients will require optimization for different column chemistries, lengths, and particle sizes.

3.4 MS Settings

Example MS settings, including cone and capillary voltage, are shown in Table 2 below. Note that these settings are for Waters Xevo Q-ToF mass spectrometers and will require optimization for each specific mass spectrometer. However, they can be used as guidelines, based on an ultra-performance liquid chromatography (UPLC) setup with a flow rate of 400–500 μL/min and a Q-ToF mass spectrometer.

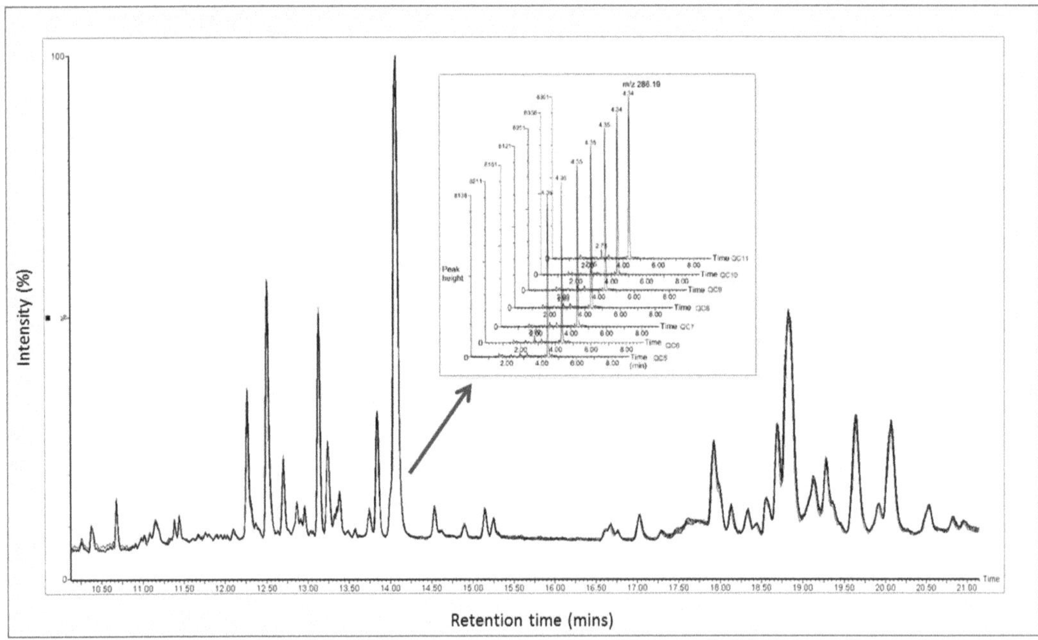

Fig. 2 Selected portion of a BPI chromatogram showing serum QC samples overlaid. Inset is an example of monitoring ion intensity of a specific metabolite over the QC samples

Table 1a

Example gradient for the *RP analysis of urine samples*

Time (min)	A (%)	B (%)	Comment
0	99	1	
1	99	1	
3	85	15	
6	50	50	
9	5	95	
10	5	95	Column washing
10.1	99	1	Column re-equilibration
12	99	1	

Column = UPLC Acquity HSS T3. Flow rate = 0.5 mL/min. Mobile phases; A = 0.1% formic acid in water, B = 0.1% formic acid in acetonitrile. *Taken from Ref.* 21

3.5 MS/MS Experiments

DDA experiments—data-dependent or data-directed experiments: Tandem mass spectrometry experiments can be set up, where a specific number of precursor ions are selected using predetermined rules and thresholds, such as exceeding a certain ion intensity.

Table 1b

Example gradient for the *HILIC analysis of urine samples*

Time (min)	A (%)	B (%)	Comment
0	99	1	
1	99	1	
12	0	100	Column washing
12.1	99	1	Column re-equilibration
15	99	1	

Column = UPLC Acquity HILIC. Flow rate = 0.4 mL/min. Mobile phases; A = 95% acetonitrile, 5% water + ammonium acetate (10 mM final concentration), B = 50% acetonitrile, 50% water + ammonium acetate (10 mM final concentration). *Taken from reference* 21

Table 1c

Example gradient for *the RP analysis of plasma/serum samples*

Time (min)	A (%)	B (%)	Comment
0	99.9	0.1	
2	99.9	0.1	
6	75	25	
10	20	80	
12	10	90	
21	0.1	99.9	Column washing
23	0.1	99.9	
24	99.9	0.1	Column re-equilibration
26	99.9	0.1	

Column = UPLC Acquity HSS T3. Flow rate = 0.4 mL/min. Mobile phases; A = 0.1% formic acid in water, B = 0.1% formic acid in methanol. *Taken from reference* 22

These MS/MS experiments can provide information which may aid in the structural identification of discriminatory metabolites. Although it is often the case that further MS/MS experiments need to be performed, once such potential biomarkers have been selected from data analysis, performing DDA experiments at this time does not impact negatively on the study setup.

Table 1d

Example gradient for the *HILIC analysis of plasma/serum samples and aqueous tissue extracts*

Time (min)	A (%)	B (%)	Comment
0	99	1	
2	99	1	
8	45	55	
9	1	99	
9.1	1	99	Flow rate increased to 0.8 mL/min
11	1	99	Column washing
11.1	99	1	Column re-equilibration
19	99	1	
19.1	99	1	Flow rate decreased to 0.4 mL/min
23	99	1	

Flow rate = 0.4 mL/min unless otherwise stated. Column = UPLC Acquity HILIC. Mobile phases; A = acetonitrile/water (95:5), B = acetonitrile/water (50:50). *Taken from reference* 25

Table 1e

Example gradient for the *RP analysis of tissue samples—aqueous extract*

Time (min)	A (%)	B (%)	Comment
0	99.9	0.1	
2	99.9	0.1	
6	75	25	
10	20	80	
12	10	90	
21	0.1	99.9	Column washing
23	0.1	99.9	
24	99.9	0.1	Column re-equilibration
26	99.9	0.1	

Column = UPLC Acquity HSS T3. Flow rate = 0.4 mL/min. Mobile phases; A = 0.1% formic acid in water, B = 0.1% formic acid in methanol. *Taken from reference* 23

Table 1f

Example gradient for the *RP analysis of tissue samples—organic extract*

Time (min)	A (%)	B (%)	Comment
0	60	40	
2	57	43	
2.1	50	50	
12	46	54	
12.1	30	70	
18	1.0	99	Column washing
18.1	60	40	Column re-equilibration
20	60	40	

Column = UPLC Acquity CSH. Flow rate = 0.4 mL/min. Mobile phases; A = acetonitrile (ACN)/water (60:40); B = isopropanol/ACN (90:10). In both mobile phases ammonium formate was diluted to 10 mM and formic acid to 0.1%. *Taken from reference* 25

Table 1g

Example gradient for the *HILIC analysis of tissue samples—aqueous extracts*

Time (min)	A (%)	B (%)	Comment
0	99	1	
2	99	1	
8	45	55	
9	1	99	
9.1	1	99	0.8 mL/min column washing
11	1	99	0.8 mL/min column washing
11.1	99	1	0.8 mL/min column washing
19	99	1	Column re-equilibration
23	99	1	

Column = Acquity HILIC. Temperature = 35 °C. Flow rate = 0.4 mL/min unless otherwise stated. Mobile phases; A = acetonitrile (ACN)/water (95:5); B = ACN/water (50:5). In both A and B, the concentration of ammonium acetate is 10 mM and formic acid is present at 1%. *Taken from reference* 25

Table 2

Example MS settings for positive and negative mode ESI analysis using a Q-ToF mass spectrometer. The main parameters are shown. These will need to be optimized for each instrument and ESI mode and are to some extent dependent on solvent system and mobile phase flow rate

Parameter	Setting
Capillary voltage	1–3 kV electrospray (ESI)+,1–2.5kVESImode-
Cone voltage	30 V
Source temperature	e.g., 120 °C
Desolvation temperature	e.g., 350 °C
Cone gas flow	25 L/h
Desolvation gas flow	900 L/h

3.6 Data Preprocessing

It is beyond the scope of this protocol to provide a detailed description of data analysis. However, the key data analysis steps pertaining to LC-MS untargeted analysis for metabolic profiling studies are described briefly in **steps 1–7**. There are many different commercial and freeware available to perform some or all of these steps. These are listed in the Materials section of this protocol.

1. The first step in data preprocessing of untargeted LC-MS data is peak picking. Define chromatographic processing regions—it may be that it is not desirable to include the solvent front, e.g., 0–1 min, and the re-equilibration portion of the data. Some software may allow the user to omit these regions from further analysis.

2. This is followed by alignment of the peaks to correct for any retention shifts that have occurred between samples during the analysis.

3. Peaks are then integrated using peak area through a selected algorithm depending on the software used.

4. Data is often normalized during this process. A common approach is to use median fold change [26] to account for differences in sample dilutions. These are particularly common in biological samples such as urine samples.

5. QC filtering is key in untargeted LC-MS metabolic profiling studies. This is explained in Subheading 3.7.

6. Output of metabolite feature table for multivariate analysis.

3.7 Data Analysis

Once the metabolite table has been produced, this can be exported into appropriate software, e.g., excel for further analysis, e.g., QC CV filtering. This will be covered in more detail in Chapter 2.

3.8 Database Searching for Structural Identification

Still a bottleneck in LC-MS untargeted analysis, structural identification is key in metabolic profiling studies. Databases such as HMDB (http://www.hmdb.ca/), METLIN (https://metlin.scripps.edu/index.php), and LIPIDMAPS (http://www.lipidmaps.org/) are growing in size as research groups and individual researchers add to them. These databases can be searched in a few ways in order to attempt to identify candidate biomarkers, such as MS searches, MS/MS searches, or text searches. Readers are directed to reviews in this area [27, 28].

The standard way to proceed with structural identification is to carry out MS/MS experiments in order to produce fragmentation patterns. This will provide additional high mass accuracy information to that obtained from MS spectra alone. These MS/MS spectra can be used to search databases and will enable the list of potential candidates to be narrowed significantly from solely using MS data. Untargeted MS/MS data (e.g., DDA experiments) can be collected on QC conditioning samples or any QC sample injected throughout the run, although it would be recommended to use a sample before or after the main study sample analysis. Targeted MS/MS is usually performed on selected samples containing the metabolite feature(s) of interest in high abundance. These targeted assays would be performed after univariate or multivariate analysis has been performed and potential biomarkers selected. Many of the commonly used databases allow for uploading of MS/MS fragment ion information, e.g., HMDB and METLIN. HMDB will output predicted MS/MS spectra which can be visualized against the experimental MS/MS spectra obtained during a study.

4 Notes

1. Exercise caution when preparing samples from humans, as there is a risk of infection. Samples should be handled and prepared according to appropriate biosafety protocols.

2. Tissues should be collected onto ice and snap frozen as soon as possible after collection or processed for analysis immediately. This is to avoid further metabolism occurring or metabolite losses.

3. Note that tissue samples can be prone to contamination from sources such as anesthetics, surgical tubes, or instruments, e.g., cleaning solutions.

4. Tissue samples should be stored at the lowest possible temperature, ideally −80 °C but otherwise −40 °C. This will help to minimize further metabolite losses.

5. At the time of collection, sub-aliquot samples, including tissue samples, are placed into appropriate storage containers to avoid freeze-thaw issues.

6. Exercise caution when handling solvents for sample preparation and the preparation of mobile phases.

7. For the extraction of tissue samples, the volume of extraction solvent stated in this protocol was optimized for 50 mg (+/−5 mg) of tissue. This volume can therefore be scaled down if the amount of tissue is less. It must be noted that the bead-beater tubes or Eppendorf tubes are 2 mL, and so if >1.5 mL of liquid is placed into them along with the tissue sample, there is a risk of leaking through the lid of the tube during the extraction process.

8. The number and speed of the bead beating cycles can be adjusted according to the nature of the tissue. For example, fibrous tissue such as veins and placental samples may need more cycles operated at a higher frequency, e.g., three cycles at 6500 rather than 4000 mHz.

9. There is a risk with the addition of stainless steel beads to organic solvents in Eppendorf tubes that the tubes can degrade during the tissue lysis process, resulting in damage and/or sample leakage.

10. When preparing the mobile phases, it is important to note that the addition of salts, particularly to mobile phases with a high organic content, will require more careful preparation. This is the case for both HILIC and lipid mobile phases. The key to successful preparation is to:

    (a) Prepare in advance, e.g., the day before sample analysis commences.

    (b) Sonicate the bottles containing the mobile phases during preparation to ensure salts have dissolved, ideally at >40 °C. Allow the mobile phases to sit at room temperature for at least 1 h before use; check no precipitation has occurred.

11. QC Samples. Alternatively, or perhaps in some cases, additionally, a commercially purchased external QC sample could be used. This has been used in the case of serum[15]. Although this sample would not be representative of the study samples, it offers value in terms of conditioning the column and assessing instrument drift.

12. Column temperature must be adjusted depending on the mobile phase composition—mixtures with methanol results in

higher viscosity than when using acetonitrile and therefore higher backpressure. Increasing column temperature, e.g., to 50–55 °C, will help to reduce column backpressure.

13. When storing and analyzing organic extracts, 96-well plates are best avoided as there is the potential of contaminants leaching into the solvent and affecting the analysis. It is advised to store the samples in glass MS vials.

14. When using MS vials, where possible, use vials from the same batch (there are 100 vials in a box) as there may be differences between the batches in terms of purity, which can affect the analysis.

15. With larger studies, e.g., >100 samples, the analytical run time is likely to be more than 24 h. Therefore, samples may be in the autosampler at 4 °C for up to a few days, resulting in potential degradation of less stable metabolites. It is important that sample stability is assessed during analysis through the use of quality control samples and perhaps also the system suitability or test mix.

References

1. Holmes E, Loo RL, Stamler J et al (2008) Human metabolic phenotype diversity and its association with diet and blood pressure. Nature 453(7193):396–400. https://doi.org/10.1038/nature06882

2. Newgard CB (2017) Metabolomics and metabolic diseases: where do we stand? Cell Metab 25(1):43–56. https://doi.org/10.1016/j.cmet.2016.09.018

3. Brennan L (2016) Metabolomics in nutrition research-a powerful window into nutritional metabolism. Essays Biochem 60(5):451–458

4. Saurina J, Sentellas S (2017) Strategies for metabolite profiling based on liquid chromatography. J Chromatogr B 1044-1045:103–111. https://doi.org/10.1016/j.jchromb.2017.01.011

5. Haggarty J, Burgess KE (2016) Recent advances in liquid and gas chromatography methodology for extending coverage of the metabolome. Curr Opin Biotechnol 43:77–85. https://doi.org/10.1016/j.copbio.2016.09.006.

6. Markley JL, Brüschweiler R, Edison AS et al (2016) The future of NMR-based metabolomics. Curr Opin Biotechnol 43:34–40. https://doi.org/10.1016/j.copbio.2016.08.001

7. Chen Y, Xu J, Zhang R, Abliz Z (2016) Methods used to increase the comprehensive coverage of urinary and plasma metabolomes by MS. Bioanalysis 8(9):981–997. https://doi.org/10.4155/bio-2015-0010

8. Wilson ID, Nicholson JK, Castro-Perez J et al (2005) High resolution "ultra performance" liquid chromatography coupled to oa-TOF mass spectrometry as a tool for differential metabolic pathway profiling in functional genomic studies. J Proteome Res 4(2):591–598

9. Nassar AF, Wu T, Nassar SF, Wisnewski AV (2017) UPLC-MS for metabolomics: a giant step forward in support of pharmaceutical research. Drug Discov Today 22(2):463–470. https://doi.org/10.1016/j.drudis.2016.11.020

10. Zhao YY, Lin RC (2014) UPLC-MS(E) application in disease biomarker discovery: the discoveries in proteomics to metabolomics. Chem Biol Interact 215:7–16. https://doi.org/10.1016/j.cbi.2014.02.014

11. Wang X, Sun H, Zhang A (2011) Ultra-performance liquid chromatography coupled to mass spectrometry as a sensitive and powerful technology for metabolomic studies. J Sep Sci 34(24):3451–3459. https://doi.org/10.1002/jssc.201100333

12. Siskos AP, Jain P, Römisch-Margl W et al (2016) Interlaboratory reproducibility of a targeted metabolomics platform for analysis of human serum and plasma. Anal Chem 89(1):656–665

13. Michopoulos F, Whalley N, Theodoridis G et al (2014) Targeted profiling of polar intracellular metabolites using ion-pair-high performance liquid chromatography and -ultra high performance liquid chromatography coupled to tandem mass spectrometry: applications to serum, urine and tissue extracts. J Chromatogr A 1349:60–68. https://doi.org/10.1016/j.chroma.2014.05.019

14. Monteiro MS, Carvalho M, Bastos ML et al (2013) Metabolomics analysis for biomarker discovery: advances and challenges. Curr Med Chem 20(2):257–271

15. Dunn WB, Broadhurst D, Begley P et al (2011) Human serum metabolome (HUSERMET) consortium. Procedures for large-scale metabolic profiling of serum and plasma using gas chromatography and liquid chromatography coupled to mass spectrometry. Nat Protoc 6(7):1060–1083. https://doi.org/10.1038/nprot.2011.335.

16. Gray N, Lewis MR, Plumb RS et al (2015) High-throughput microbore UPLC-MS metabolic phenotyping of urine for large-scale epidemiology studies. J Proteome Res 14(6):2714–2721. https://doi.org/10.1021/acs.jproteome.5b00203

17. Wilson ID (2015) Metabolic phenotyping by liquid chromatography-mass spectrometry to study human health and disease. Anal Chem 87(5):2519. https://doi.org/10.1021/acs.analchem.5b00409. No abstract available

18. Vorkas PA, Isaac G, Anwar MA et al (2015) Untargeted UPLC-MS profiling pipeline to expand tissue metabolome coverage: application to cardiovascular disease. Anal Chem 87(8):4184–4193. https://doi.org/10.1021/ac503775m

19. Gika HG, Theodoridis GA, Plumb RS et al (2014) Current practice of liquid chromatography-mass spectrometry in metabolomics and metabonomics. J Pharm Biomed Anal 87:12–25. https://doi.org/10.1016/j.jpba.2013.06.032

20. Spagou K, Wilson ID, Masson P et al (2011) HILIC-UPLC-MS for exploratory urinary metabolic profiling in toxicological studies. Anal Chem 83(1):382–390. https://doi.org/10.1021/ac102523q

21. Virgiliou C, Sampsonidis I, Gika HG et al (2015) Development and validation of a HILIC- MS/MS multi-targeted method for metabolomics applications. Electrophoresis 36:2215–2225

22. Want EJ, Wilson ID, Gika H et al (2010) Global metabolic profiling procedures for urine using UPLC-MS. Nat Protoc 5(6):1005–1018. https://doi.org/10.1038/nprot.2010.50

23. Want EJ, Coen M, Masson P et al (2010) Ultra performance liquid chromatography-mass spectrometry profiling of bile acid metabolites in biofluids: application to experimental toxicology studies. Anal Chem 82(12):5282–5289. https://doi.org/10.1021/ac1007078

24. Want EJ, Masson P, Michopoulos F et al (2013) Global metabolic profiling of animal and human tissues via UPLC-MS. Nat Protoc 8(1):17–32. https://doi.org/10.1038/nprot.2012.135.

25. Vorkas PA, Shalhoub J, Isaac G et al (2015) Metabolic phenotyping of atherosclerotic plaques reveals latent associations between free cholesterol and ceramide metabolism in atherogenesis. J Proteome Res 14(3):1389–1399. https://doi.org/10.1021/pr5009898

26. Veselkov KA, Vingara LK, Masson P et al (2011) Optimized preprocessing of ultra-performance liquid chromatography/mass spectrometry urinary metabolic profiles for improved information recovery. Anal Chem 83(15):5864–5872. https://doi.org/10.1021/ac201065j

27. Dias DA, Jones OA, Beale DJ, (2016) Current and future perspectives on the structural identification of small molecules in biological systems. Metabolites. 6(4). pii: E46.

28. Bocker S (2016) Searching molecular structure databases using tandem MS data: are we there yet? Curr Opin Chem Biol 36:1–6. https://doi.org/10.1016/j.cbpa.2016.12.010.

Chapter 8

NMR-Based Metabolic Profiling Procedures for Biofluids and Cell and Tissue Extracts

Dimitra Benaki and Emmanuel Mikros

Abstract

Metabolomic studies offer a wealth of information on cells, tissues, and biofluids. The phenotype representation through the metabolic profiling is a valuable tool for direct diagnosis, therapeutic strategies, and system's biology studies. Nuclear magnetic resonance (NMR) spectroscopy provides a nondestructive and extremely reproducible method allowing simultaneous detection of a large number of known and unknown chemical substances.

Sample collection and preparation and experimental conditions are critical for the reliability of the subsequent analysis. The pre-analytical phase is decisive as it could generate biased spectral data misleading the following analysis. The formulation of standard operating procedures is thus of crucial importance in order to access meaningful samples and results. In this protocol, we provide standardized operations and routine procedures from sample preparation to determine the measurement details for the acquisition of NMR spectra highlighting major methodological issues.

Key words NMR spectroscopy, Metabolic profiling, Biofluids, Cell cultures, Tissues

1 Introduction

Metabolic profiling offers a comprehensive exploration of the metabolome in biological systems encompassing advanced analytical techniques and multivariate statistical analyses. Solution-state nuclear magnetic resonance (NMR) spectroscopy is a proven and versatile tool of choice to simultaneously detect, in a rapid and reproducible way the constituents of complex mixtures like biological samples.

NMR spectroscopy-based metabolic profiling provides a holistic view of the alterations of a living system, monitoring in an untargeted manner a variety of low molecular weight (MW) metabolites, fluctuating under pathophysiological conditions or genetic modifications. NMR-based metabolomics is applied in disease diagnosis, toxicology, system's biology, and functional genomics, as well as in studies that investigate how living organisms

Georgios A. Theodoridis et al. (eds.), *Metabolic Profiling: Methods and Protocols*, Methods in Molecular Biology, vol. 1738,
https://doi.org/10.1007/978-1-4939-7643-0_8, © Springer Science+Business Media, LLC, part of Springer Nature 2018

interact with their environment. Metabolites can be considered as the terminal endpoint of biochemical processes influenced by various external factors like diet (nutrimetabolomics), therapeutic intervention (pharmacometabolomics), lifestyle, or others. A wide range of biological samples such as urine, plasma, serum, cerebrospinal fluid (CSF), saliva, synovial fluid, semen, exhaled breath condensate (EBC), as well as cell and tissue (organ, tumor, muscle) homogenates can be analyzed with high-throughput automatic routines coupled with powerful data analysis methods. The access to metabolomic analysis along with other -omic methodologies like genomics, transcriptomics, and proteomics has been characterized to be the "tour de force no. 1" of personalized medicine in the future [1].

The main advantages of NMR spectroscopy for metabolic profiling of biological samples are:

1. Rapid information recovery in a nondestructive manner. One-dimensional NMR spectra can be acquired in a few minutes, and the sample is fully recovered and can be reused.

2. Small sample volumes (about 500 mL).

3. Minimum sample preparation.

4. No pretreatment is required for samples like blood, plasma/serumor urine.

5. Simultaneous detection of large number of small MW metabolites.

6. No need for preselection of analytical conditions as required in the chromatographic techniques, where specific experimental parameters (column, detector, solvent) are required.

7. Detection over an extended concentration range.

8. Excellent reproducibility that makes ^{1}H NMR spectroscope as the method of choice for large epidemiological studies.

9. Detection of metabolites with no restrictions relating to volatility, polarity, and the presence of specific chromophores.

The main limitation of NMR is sensitivity as the detection limit is in the micromolar range. However, higher magnetic fields allow for better sensitivity and resolution of NMR spectra, while the use of modern cryogenic probe heads (cryoprobes) can increase the sensitivity several fold. Furthermore, integration of the system with robotic sample changers and increased stability of conditions during measurements (e.g., temperature) provide high accuracy and reproducibility. Finally, NMR can be hyphenated with chromatographic techniques and provide structural information complementary to MS. The most common analysis in NMR-based metabolomics consists of simple 1D ^{1}H spectra that can be acquired in a few minutes per sample; however, in complex mixtures, 2D homonuclear and heteronuclear NMR can be employed giving the possibility to identify unknown metabolites in an accurate manner.

NMR is very sensitive to the pre-analytical processing of the biosamples (collection, transportation, storage, sample preparation). In the case of biofluid collections, the daily routine of clinical facilities may impose major obstacles. Well-standardized operations of the pre-analytical phase are prerequisite for the reliability of the following analysis. It is thus of crucial importance the formulation of feasible SOPs in order to access meaningful samples and results.

2   Materials

2.1 Sample Collection and Metabolite Extraction

1. Phosphate-buffered saline (PBS).

2. Ultrapure water.

3. Chloroform, analytical grade, $\geq$99%.

4. Methanol, LC-MS grade, $\geq$99.8%.

   Dry ice and liquid N_2 are required for the extraction procedure and bovine serum albumin (BSA) and 1 M NaOH for the protein content determination.

2.2 NMR Sample Preparation

1. KH_2PO_4 or $Na_2HPO_4 \times 7H_2O$ or Na_2HPO_4 anhydrous >99%.

2. KOH pellets or HCl.

3. Sodium azide (NaN_3).

4. 3-(Trimethylsilyl)propionic-2,2,3,3-D_4 acid sodium salt (TSP), >98 atom % D.

5. Tetramethylsilane (TMS), >99%.

6. D_2O, 99.9 atom % D.

7. $CDCl_3$, 99.8% D.

2.3 Laboratory Equipment

1. Cell scraper, 18 mm blade.

2. High precision pipettes.

3. Eppendorf tubes, 1.5 mL.

4. Centrifuge tubes, 15 mL.

5. Tissue homogenizer or mortar and pestle.

6. Lowry assay.

7. Refrigerated centrifuge.

8. Ultrasound bath.

9. Sample freeze drier or centrifugal vacuum concentrator.

10. −80 °C freezer.

11. Heparin blood collection tubes for blood sample collection.

12. Sterile urine collection tubes.

2.4 Instrumentation: NMR Equipment

1. 600 MHz NMR spectrometer is advisable for biological samples as a compromise in terms of resolution and sensitivity vs cost.

2. Probe of 5 mm diameter with Z gradient (e.g., BBI, TXI for Bruker Biospin).

3. Temperature control unit; N_2 supply provides higher stability compared to dry air.

4. Automatic sample changer (B-ACS60, Bruker Biospin, or similar).

5. Software to control sample loading, temperature equilibration and stability, matching and tuning, shimming, pulse calibration, acquisition, and processing (ICONNMR, Bruker Biospin, or similar).

2.5 Buffers Preparation

Buffer A (50 mL): weigh 10.2 g KH_2PO_4, 50 mg TSP, and 6.5 mg NaN_3, add 40 mL D_2O, and dilute in ultrasonic bath. Adjust the pH to 7.4 with KOH pellets (for a volume of 50 mL around 33–35 pellets are required). Fill up with D_2O to 50 mL volume and mix very well. Aliquot and store at 4 °C.

Buffer B (500 mL): weigh 10.05 g $Na_2HPO_4 \times 7H_2O$ (or 5.32 g Na_2HPO_4), 0.4 g TSP, and 0.2 g NaN_3, add 380 mL ultrapure water, and dilute in ultrasonic bath. Adjust the pH at 7.4 with 1 M HCl. Fill volume up to 400 mL with ultrapure water. Add 100 mL D_2O and mix well. Aliquot and store at 4 °C.

Check always the pH before use.

3   Methods

3.1 Sample Collection

3.5.1 Cells

The initial treatment step on the biological system has to be acute and effective (to stop all the enzymatic activity) in order to freeze the existing state as well as to ensure system integrity and to generate reproducible results.

It is generally accepted that approximately 10^7 cells produce NMR samples of the necessary quality for metabolomic studies.

Sample collection in monolayer cultures

1. Aspirate the cell medium or collect and store at −80 °C for further analysis of the excreted metabolites (*see* **Note 1**).

2. Place the culture dishes on ice and wash twice with ice-cold PBS to remove residual traces of culture medium.

3. Add ice-cold methanol up to the necessary volume to cover the culture surface (2 mL for the 10-cm-diameter culture dish, 4 mL for 75 cm^2 flask or adjust the volume according to the

culture vessel). Immediate quench of the cell metabolism is decisive for the quality of the study.

4. Leave on ice for 5 min.

5. Use a scraper for the mechanical detachment of the cells (*see* **Note 2**).

6. Transfer the suspension into a 15 mL centrifuge tube and proceed to the metabolite extraction (*see* Subheading 3.2.1.). To store the cells, ice-cold PBS is used for cell quenching instead of methanol (**step 3**). The cell quench is achieved solely by the low temperature because methanol causes cell membrane disruption and consequent intracellular metabolite leakage [2]. The suspension (PBS and cells) is centrifuged at $230 \times g$ for 5 min, the supernatant is aspirated, and the cell pellet is stored at −80 °C till the metabolite extraction procedure.

Sample collection in cell suspension

1. Aliquot the desired number of cells (around 10^7) into a clean 15 mL centrifuge tube, and centrifuge at $230 \times g$ for 5 min.

2. Discard the supernatant or store at −80 °C for further analysis.

3. Suspend the cells twice in ice-cold PBS (centrifuged at $230 \times g$ for 5 min, discard, and repeat).

4. Store the cell pellet at −80 °C or resuspend the cells in 2 mL of ice-cold methanol and proceed to the metabolite extraction.

3.1.1 Tissues

Tissue Collection

General issues: (1) animals should be handled humanely in accordance with the European Union Directive 86/609 (or updated). (2) The animals should be lightly anesthetized before sacrifice with compounds that do not affect organism-organ metabolome. (3) All anesthetic agents must be listed on an approved Animal Protocol (IACUC Guidelines). (4) Treatment with compounds that are detectable through NMR (e.g., ether anesthesia for animal models) or that induce metabolic modifications [3, 4] should be avoided. A recently published study on male C57BL/6J mice using untargeted and targeted metabolomics reveals dramatic tissue-specific impacts of various collection strategies [5]. (5) Organs and tissues exhibit topological heterogeneity, e.g., ischemic and nonischemic areas in the heart and necrotic and oxygenated regions in tumors. When part of the organ will be subjected to metabolomic analysis, the same topological region should be used across the entire study. (6) Use gloves and alcohol swabs to clean the instruments after each operation.

1. Tissue collection has to be accomplished in a short time period.

2. Rinse the organs promptly with ultrapure water to remove blood in order to avoid sample contamination.

3. Dab dry and proceed to the tissue dissection in order to avoid degradation.

4. Place the tissue in prechilled, labeled containers, and immerse directly in liquid N_2.

5. Stored at −80 °C.

Tissue Homogenization

Blenders, beads shakers, or cryogenic mortar and pestle are used for tissue homogenization. Regardless of the homogenization method, all the accessories are prechilled in order to avoid tissue degradation and have to be cleaned between samples to avoid contaminations (*see* **Note 3**).

The following procedure is described for homogenization using acryogenic mortar and pestle.

1. Store the mortar and pestle overnight at −80 °C.

2. Place the mortar on dry ice along the whole procedure.

3. Poor liquid N_2 into the mortar and transfer the frozen tissue.

4. At the end of the homogenization, allow the excess of liquid N_2 to evaporate (continuous mixing).

5. Use a cold spatula (kept chilled on dry ice) to transfer the pulverized tissue promptly into chilled, pre-weighed, and labeled eppendorfs or cryovials (*see* **Note 4**).

6. Reweigh to determine the tissue mass.

7. Store at −80 °C till the extraction procedure (*see* **Note 5**).

3.1.2 Biofluids

Biofluid collection is easier than other biological samples; however, processing and time intervals are crucial for system integrity and consequent sample variability.

Blood collection is usually performed in prefilled vials with anticoagulant supplements as EDTA, citrate, or heparin. EDTA shows strong and broad resonance lines that obscure extended and important areas of the ^{1}H NMR spectra of the samples [6]. The use of non-deuterated citrate can change the endogenous metabolite concentration and/or saturate NMR receiver.

Plasma collection

1. Collect whole blood into heparin tubes in the morning preprandially, after overnight fasting.

2. Centrifuge at 1500 × *g* for 10 min at 4 °C to separate blood cells from plasma within 30 min of collection.

3. Collect the supernatant, aliquot, and store at −80 °C.

Serum collection

1. Allow blood to clot without anticoagulant on ice for 30 min.

2. Centrifuge at 1500 × *g* for 15 min at 4 °C.

3. Collect the serum, aliquot, and store at −80 °C.

   Both procedures (plasma and serum collection) should not exceed 2 h to minimize any metabolite degradation.
   Urine collection

1. Collect midstream urine in sterile containers, after overnight fasting.

2. Keep on ice or at 4 °C.

3. Centrifuged at 1850 × *g* for 10 min at 4 °C to remove debris.

4. Collect the supernatant, aliquot, and store at −80 °C.

3.2 Metabolite Extraction

To identify different sources of contaminations: (1) check the purity of the solvents to be used for the extraction procedure with NMR, i.e., reduce to dryness the same, or better double the volume, to be used and record an NMR spectrum in the corresponding deuterated solvents (methanol, chloroform, deionized water); and (2) use blank extraction samples and record NMR spectra as for the original samples [7].

In the case of the cell endo-metabolome (intracellular metabolites) and tissue homogenates, a metabolite extraction step is required to separate polar from nonpolar metabolites and isolate insoluble macromolecular entities (proteins and disrupted membranes). High MW biomolecules introduce broad, unresolvable, signals into the ^{1}H NMR spectra, disrupt the flat baseline, and make the integration of metabolite signals and comparison impossible. The three-solvent (methanol-water-chloroform) extraction protocol is widely used both in cell and tissue metabolomic studies [7–9].

3.2.1 Cell Extraction

1. For cell suspensions quenched by 2 mL ice-cold methanol, an equal volume of ice-cold chloroform is added (methanol/chloroform 1:1, adjust accordingly).

2. Sonicate for 5 min (keep the temperature of the bath low by adding ice).

3. Add 1.8 mL (adjust accordingly) of cold ultrapure water.

4. Vortex for 1 min to create a homogenous emulsion.

5. Keep on ice for 15 min (helps the separation of the two phases).

6. Centrifuge at 18,900 × *g* for 20 min at 4 °C.

7. Transfer the two layers into clean tubes, and store at 4 °C till the second round of extraction is accomplished (**step 8**).

8. Repeat the extraction procedure (add methanol/chloroform/water 2:2:1.8 and go to **step 4**).

9. Collect the two phases and pool with the first extraction fractions.

10. Dry the extracts overnight in a vacuum centrifuge or under a nitrogen stream and store at −80 °C.

3.2.2 Tissue Extraction

In the case of cell pellets stored at −80 °C, the samples are left to thaw on ice, add 2 mL ice-cold methanol, vortex to disrupt the pellet, and follow the extraction procedure described above.

1. Add ice-cold $CHCl_3$/MeOH solution (2:1, 1 mL total volume/100 mg tissue) to frozen pulverized tissue.

2. Mix vigorously for the cell enzymatic function deactivation. Use a needle to disrupt any remaining large frozen formations.

3. Sonicate in ice-water bath for 5 min.

4. Add an equivalent volume (1 mL) of cold ultrapure water.

5. Vortex vigorously for 1 min.

6. Leave on ice for 15 min.

7. Centrifuge at 18,900 × g for 20 min at 4 °C (*see* **Note 6**).

8. Collect the two phases in clean cryovials, and store at 4 °C till the second extraction run (**step 9**).

9. Repeat the extraction procedure from **step 1**.

10. Collect the two phases and pool with the first extraction fractions.

11. Dry overnight in a vacuum centrifuge or in N_2 stream and store at −80 °C.

When different tissue weights are accessible, adjust the required solvent volumes in order to generate samples of similar concentration. The collection of equal volumes from each sample extracts (**step 8**) makes easier the NMR sample preparation. For additional normalization, dry the insoluble residual formed in the biphasic interface, weigh, and solubilize 2 mL/g tissue (or in the case of cells 200 µL/pellet) to determine the protein of content using a Lowry assay [10].

3.3 NMR Sample Preparation

The use of a buffer solution is necessary to ensure the stable and constant pH of the samples. Proton resonances can be strongly dependet on the pH and small fluctuations result in signal shifts, hampering further analysis. The buffers used in NMR studies are the sodium or potassium phosphates, as their capacity covers the physiological pH range. It is advisable to generate quality control (QC) samples by pooling 5 µL of each sample.

3.3.1 Cell and Tissue Extracts: Polar Metabolites

The dried cell and tissue extracts are reconstituted in 10% buffer A and 90% D_2O (*see* **Note 7**).

1. Prepare the necessary volume of 10% buffer A in D_2O 99.9%.

2. Leave the samples to thaw at room temperature.

3. Add 650 μL of the solution (**step 1**), and mix properly (use vortex and/or ultrasonic bath) to ensure the resuspension of all the contents.

4. Centrifuge at $18,900 \times g$ for 10 min at 4 °C.

5. Transfer 550 μL of the supernatant to a clean 5-mm-diameter NMR tube.

3.3.2 Cell and Tissue Extracts: Nonpolar Metabolites

1. Prepare the necessary volume of $CDCl_3$ 0.03% v/v TMS solution. During sample reconstitution, place the bottle on ice, and close the cup immediately after use to prevent, as much as possible, TMS evaporation.

2. Leave the samples to thaw at room temperature.

3. Add 650 μL of the solution (**step 1**), and mix properly (use vortex and/or ultrasonic bath) to ensure the resuspension of all the content.

4. Centrifuge at $18,900 \times g$ for 10 min at 4 °C.

5. Transfer 550 μL of the supernatant in a clean 5-mm-diameter NMR tube.

3.3.3 Biofluids

For urine samples, buffer A is used.

1. Leave the samples to thaw on ice.

2. Transfer 630 μL in an eppendorf and add 70 μL buffer A.

3. Vortexed for 60 s.

4. Centrifuge at $18,900 \times g$ for 10 min at 4 °C.

5. Transfer 550 μL of the supernatant in a clean 5-mm-diameter NMR tube.

For other biofluids as EBC and CSF, as well as the extracellular cell culture medium, the NMR samples are prepared as urine samples. Additional precautions should be taken in the case of CSF samples as exposure to the atmosphere causes CO_2 evaporation, and pH can reach high values (9.0–9.7). Phosphates do not have any buffering capacity at this pH range, and solutions conditioning can be aided by the addition of DCl to lower the pH to around 7 and then add 10% buffer A.

Blood plasma and serum contain proteins and should be treated with care. Buffer B is added up to 50% of the total volume, no vortex is applied, and mixing is achieved through gentle pipetting (*see* **Note 8**).

1. Leave the samples to thaw on ice.

2. Remove particulates using a needle.

3. Transfer 350 μL to an eppendorf and add 350 μL buffer B and mix very gently (pipette up and down).

4. Transfer 550 μL in a clean 5-mm-diameter NMR tube.

3.4 NMR Experiments and Parameters

A major challenge of fundamental importance in NMR metabolomics spectra, is to remove water resonance (or HDO signals in the case of cell and tissue extract solutions) in a smooth and effective way in order not to affect the nearby resonances.

Another important issue is the presence of large MW entities, as such proteins and lipids, in certain biofluids (plasma, serum). Molecules with short tumbling time give rise to high-intensity broad signals in the ^{1}H NMR spectra that obscure the analysis of small MW entities. A pulse sequence that filters proton resonances according to T2 relaxation rate is the Carr-Purcell-Meiboom-Gill (CPMG) resulting in a flat baseline proton spectrum. The CPMG pulse sequence offers the opportunity to depict resolved signals of the low MW metabolites without discarding the proteinaceous content. Protein content transfers valuable information (e.g., LDL and HDL relative concentration in blood -derived samples) which can be accessed through a diffusion-editing pulse sequence with bipolar-gradient pulses that allow obtaining a reasonable amount of diffusion of the lipoprotein signals. Alternatively, the proteins can be removed by an additional ultrafiltration or precipitation step [11].

Before the setup of NMR experiments:

1. Confirm the high-quality performance of the spectrometer following the control SOPs defined by the manufacturer using appropriate standard solutions.

2. Perform a temperature calibration (*see* **Note 9**).

3. Optimize field homogeneity and water presaturation (prior to biofluid samples runs) using the manufacturer's standard solution (sucrose 2 mM, 0.5 mM DSS, 2 mM NaN_3 in $H_2O/$ D_2O 90/10) (part of the SOP).

4. Randomize sample analysis.

All NMR spectral parameters are optimized for a single sample, and the automation program uses this set for all the samples of the run (*see* **Note 10**). The sample of choice can be a QC sample, being the most representative; while not a real specimen, a QC sample contains the fingerprint information from the whole sample collection. If no QC samples are available, a quite concentrated one is selected (in urine samples, the color is indicative).

Water frequency determination is crucial as this value will be used for the water signal suppression in all the spectra acquired in a run.

Relaxation delay is different for the various fluids and should be optimized according to the phase distortion of the residual water signal (*see* **Note 11**).

The number of scans in the case of biofluids as urine, plasma, and serum samples is adjusted to 32. More scans, 128, 192, or even 256, are required for extracted metabolites (cells and tissues).

Receiver gain. A parameter that is optimized and fixed is the receiver gain value (e.g., adjusted at 90.5 for 600 MHz Bruker instruments).

3.4.1 Specific Pulse Sequence Parameters

1. The proton 1D experiment, holding the major metabolomic information (with the exception of samples with high content of macromolecules), is acquired using the NOESY-presaturation pulse sequence with gradients (noesygppr1d, Bruker library) offering the optimum water suppression (*see* **Note 12**). A spectral width of 20 ppm is required with a sampling of 64k points resulting in an acquisition time of 2.7 s. The mixing time at 10 ms for the NOESY sequence is optimum in combination with the presaturation and provides the best compromise for water suppression and relaxation effect suppression thus affecting peak quantification to a lesser extent. Plasma and serum spectral width is adjusted to 30 ppm, and sampling is increased up to 96k points resulting thus in an acquisition time of 2.7 s. The 90 deg. pulse width should be optimized for each sample separately (part of the acquisition automation routine) and kept constant for all the spectra of the same sample. The processing includes zero filling and exponential multiplication, phase correction, and axis calibration.

2. *J-resolved pdeudo 2D* experiments are very fast (5–10 min, depending on the analysis and number of scans) and of utmost importance offering the possibility to resolve overlapped signals. A spectral width of 16 ppm in the ^{1}H axis, while 70–80 Hz for the J coupling, is enough, with 12k points, 40 increments, and 4–8 scans (4×n). Processing includes zero filling in both dimensions (to 16 k and 256 for F2 and F1, respectively), line broadening multiplication with a factor of 0.3 Hz, baseline correction, an additional tilt by 45 deg. step, and symmetrization about the J 0 Hz line.

    Additional spectra are required in the case of plasma and serum samples (and generally for biofluids with proteins), either to reduce the disturbance created by the macromolecular entities contribution (CPMG) or specific to derive the information carried by large biomolecules (diffusion edited).

3. The T2 filter (relaxation edited) using the *Carr-Purcell-Meiboom-Gill* sequence is applied to suppress the high MW contribution (cpmgpr1d pulse sequence, Bruker library, with

presaturation during relaxation delay) in combination with the presaturation scheme. The spectral width is set to 20 ppm, with 72k points for data sampling resulting in a 3.1 s acquisition time and 32 scans. Combination of 128 loops for the T2 filtering with spin-echo delay of 300 μs results in a total echo time of 77 ms. For the processing, zero filling is implemented adding up to 128k data points and a line broadening of 1 Hz is used prior to Fourier transformation.

4. The *diffusion-edited pulse sequence* (molecular diffusion coefficient edited, ledbpgppr2s1d pulse sequence, Bruker library, with presaturation during relaxation delay) uses a bipolar pulse pair-longitudinal eddy current delay to analyze mainly the lipid content of plasma lipoproteins. A Spectral width of 30 ppm is applied using 96k data points and 2.7 s acquisition time. Typical optimized values are diffusion delay (big delta, Δ) 120 ms, 3 ms (little delta, δ), delay for the eddy current decay 5 ms, diffusion gradient length that allows editing 1.5 ms, and spoil gradient length 600 μs. Processing includes zero filling to 131k data points and a line broadening of 1 Hz.

   In plasma and serum samples, the glucose alpha-anomeric proton resonance at 5.23 ppm is used as an internal standard for the chemical shift axis calibration, due to interaction of the TSP with protein molecules (*see* **Note 13**).

5. Simple ^{1}H 1D NMR spectrum, i.e., without solvent suppression, using an optimized 90 deg. pulse is acquired for lipid metabolite profiling. Preferentially, a 30 deg. pulse is used in order to the improve signal-to-noise ratio at the same time. Spectra are acquired with a spectral width of 20 ppm, 64k data points, 2.73 s acquisition time, and 64 transients. Usually spectra are recorded at 295 K. Processing includes zero filling to 131k data points and a line broadening of 1 Hz.

   Although J-resolved spectrum is very important for the analysis of overlapped regions of the spectra, 2D experiments are always performed on selected samples for the unambiguous identification of spin systems (TOCSY spectrum) and ^{1}H-^{13}C correlations (HSQC-DEPT edited spectrum).

6. 2D TOCSY experiments (dipsi2phpr pulse sequence, Bruker library) are recorded with a spectral width of 16 ppm, 2k time domain points, and 256 increments. Receiver gain and number of scans are adjusted according to sample concentration.

7. 2D HSQC (hsqcedetgpsisp2.3 pulse sequence, Bruker library) with multiplicity editing is recorded for a spectral width of 16 ppm for the ^{1}H dimension and 190 ppm for ^{13}C, 2k time domain points, and 256 increments. Receiver gain and the number of scans are adjusted according to the sample concentration.

4   Notes

1. Cell culture medium contains sugars and amino acids as nutrition supplements and metabolites secreted into the medium related to cellular function.

2. The use of a scraper for the mechanical detachment of cells is preferred compared to standard trypsinization to avoid the use of additional compounds. However, in a recent study, Kapoore and coworkers [12] reported that leakage of metabolites during trypsinization and scraping treatments may be dependent on cell membrane architecture, affecting the recovery of different metabolite classes for different cell lines. Based on these data, mechanical detachment is safer when the extraction procedure directly follows cell collection, and methanol with the cells will be used in the extraction. For cell storage, trypsinization can be used, but the supernatant should be checked for metabolite leakage.

3. For cleaning, blenders are disassembled, and all compartments are rinsed with deionized water and methanol, while a paper roll can be used to clean thoroughly mortar and pestle. Blenders may introduce temperature increase in the specimen depending on the time necessary for the homogenization (soft tissues as the liver, hard tissues as certain solid tumors).

4. Handle eppendorfs from the cup as any contact elevates the temperature and might cause tissue defrost. The eppendorfs are immersed in liquid N_2 before and immediately after the transfer of the ground tissue.

5. Eppendorfs containing pulverized tissue, when transferred from liquid N_2 to $-80\ ^{\circ}C$, should be left with open cups for about 1 h, to allow N_2 evaporation and avoid the development of high pressure that may spread the pulverized tissue. Tissue homogenates are stored at $-80\ ^{\circ}C$.

6. If an emulsion is still present, a 2 min centrifugation is usually enough to separate the two phases.

7. It is advisable to prepare the reconstitution solution for all the samples in the same run, in order to minimize errors of pipetting (small fluctuations of the volume affect the quality of the automated optimization of field homogeneity before spectral acquisition).

8. Pipette with care to prevent foam formation due to the proteinaceous content. Remove particulate matter with the help of a needle.

9. Cell and tissue extracts, as well as biofluids spectra, are recorded at 300 K, except for plasma and serum spectra which are recorded at 310 K. For this temperature range, the calibration

standard is deuterated methanol solution, 99.8% D, in sealed NMR tubes, ideally provided by the manufacturer.

10. For the field homogeneity, the internal standard provides the quality control. The half-height line width (HHLW) value of the TSP signal at 0.00 ppm in different biofluids can be used. In cell and tissue extracts, and in urine samples, the HHLW value should be <1 Hz, and Si satellites should be clearly visible, while in samples with protein content, such as blood plasma and serum, it is expected to be onefold higher.

11. In urine, plasma, serum, and cell extract samples, a value of 4 s is optimum, while in CSF samples and tissue extracts, a higher value (up to 10 s) is required.

12. For water suppression, the pulse sequence of choice is a combination of the soft presaturation scheme with the a 1D NOESY pulse sequence with a short mixing time, preferably 10 ms. In more recent pulse sequences, spoil gradients have been introduced. The power of the water suppression pulse is as low as necessary, in order to introduce the least possible distortion to adjacent signals. More forceful pulse schemes, such as Watergate and sculpting, although more effective, affect resonances close to the water signal, i.e., the anomeric proton of sugars, the Cα proton of several amino acids, as well as baseline quality. The pulse power for water presaturation (pl9, Bruker instruments) is 25 Hz for H_2O/D_2O samples (urine, CSF, EBS, plasma) or even less e.g., 5–10 Hz in the case of 100% D_2O samples (cell and tissue extracts).

13. In urine samples, the creatinine resonance can be used as an internal standard for quantitation purposes.

Acknowledgments

The authors wish to acknowledge Eberhard Humpfer and Manfred Spraul (Bruker BioSpin, Karlsruhe) for useful help and advice on NMR parameters optimization.

References

1. Chen R, Mias GI, Li-Pook-Than J et al (2012) Personal omics profiling reveals dynamic molecular and medical phenotypes. Cell 148(6):1293–1307. https://doi.org/10.1016/j.cell.2012.02.009. Cohen J (March, 2012). Examining his own body, stanford geneticist stops diabetes in its tracks. News.sciencemag.org. Retrieved from http://www.sciencemag.org/news/2012/03/examining-his-own-body-stanford-geneticist-stops-diabetes-its-tracks

2. Dietmair S, Timmins NE, Gray PP et al (2010) Towards quantitative metabolomics of mammalian cells: development of a metabolite extraction protocol. Anal Biochem 404:155–164. https://doi.org/10.1016/j.ab.2010.04.031

3. Collinet H, Renault D (2012) Metabolic effects of CO_2 anaesthesia in Drosophila Melanogaster. Biol Lett 8:1050–1054. https://doi.org/10.1098/rsbl.2012.0601

4. Ghini V, Unger FT, Tenori L et al (2015) Metabolomics profiling of pre-and post-anesthesia plasma samples of colorectal patients obtained via Ficoll separation. Metabolomics 11:1769–1778. https://doi.org/10.1007/s11306-015-0832-5

5. Overmyer KA, Thonusin C, Qi NR et al (2015) Impact of anesthesia and euthanasia on metabolomics of mammalian tissues: studies in a C57BL/6J mouse model. PLoS One 10(2):e0117232. https://doi.org/10.1371/journal.pone.0117232

6. Nicholson JK, Buckingham MJ, Sadler PJ (1983) High resolution ^{1}H NMR studies of vertebrate blood and plasma. Biochem J 211(3):605–615

7. Keun HC, Athersuch TJ (2011) Nuclear magnetic resonance (NMR)-based metabolomics. In: Metz TO (ed) Metabolic profiling, Methods in molecular biology, vol vol 708. Springer Protocols, Humana Press, New York, pp 321–334. https://doi.org/10.1007/978-1-61737-985-7

8. Beckonert O, Keun HC, Ebels TMD et al (2007) Metabolic profiling, metabolomic and metabonomic procedures for NMR spectroscopy of urine, plasma, serum and tissue extracts. Nat Protoc 2(11):2692–2703. https://doi.org/10.1038/nprot.2007.376

9. Sapcariu SC, Kanashova T, Weindl D et al (2014) Simultaneous extraction of proteins and metabolites from cells in culture. MethodsX 1:74–80. https://doi.org/10.1016/j.mex.2014.07.002

10. Le Belle JE, Harris NG, Williams SR et al (2002) A comparison of cell and tissue extraction techniques using high-resolution ^{1}H-NMR spectroscopy. NMR Biomed 15:37–44

11. Gowda NGA, Raftery D (2014) Quantitating metabolites in protein precipitated serum using NMR spectroscopy. Anal Chem 86(11):5433–5440. https://doi.org/10.1021/ac5005103

12. Kapoore RV, Coyle R, Staton CA et al (2015) Cell line dependence of metabolite leakage in metabolome analyses of adherent normal and cancer cell lines. Metabolomics 11:1743–1755. https://doi.org/10.1007/s11306-015-0833-4

Chapter 9

Untargeted GC-MS Metabolomics

Matthaios-Emmanouil P. Papadimitropoulos, Catherine G. Vasilopoulou, Christoniki Maga-Nteve, and Maria I. Klapa

Abstract

Untargeted metabolomics refers to the high-throughput analysis of the metabolic state of a biological system (e.g., tissue, biological fluid, cell culture) based on the concentration profile of all measurable free low molecular weight metabolites. Gas chromatography-mass spectrometry (GC-MS), being a highly sensitive and high-throughput analytical platform, has been proven a useful tool for untargeted studies of primary metabolism in a variety of applications. As an omic analysis, GC-MS metabolomics is a multistep procedure; thus, standardization of an untargeted GC-MS metabolomics protocol requires the integrated optimization of pre-analytical, analytical, and computational steps. The main difference of GC-MS metabolomics compared to other metabolomics analytical platforms, including liquid chromatography-MS, is the need for the derivatization of the metabolite extracts into volatile and thermally stable derivatives, the latter being quantified in the metabolic profiles. This analytical step requires special care in the optimization of the untargeted GC-MS metabolomics experimental protocol. Moreover, both the derivatization of the original sample and the compound fragmentation that takes place in GC-MS impose specialized GC-MS metabolomic data identification, quantification, normalization and filtering methods. In this chapter, we describe the integrated protocol of untargeted GC-MS metabolomics with both the analytical and computational steps, focusing on the GC-MS specific parts, and provide details on any sample depending differences.

Key words Untargeted metabolomics, Gas chromatography-mass spectrometry (GC-MS) metabolomics, Metabolic profiling, Metabolic network analysis, Primary metabolism

1 Introduction

In the era of systems biology, high-throughput biomolecular (omic) analyses have been used extensively in order to obtain a global perspective of the molecular physiology of biological systems. Untargeted metabolomics pursues the metabolic physiology and concerns the high-throughput analysis of the metabolic state regulation of a tissue, biological fluid, or cell culture, through the quantification of the concentration profile of all its measurable free low molecular weight metabolites, i.e., its metabolic profile, under various physiological conditions [1–3]. Thus, metabolomics should

Georgios A. Theodoridis et al. (eds.), *Metabolic Profiling: Methods and Protocols*, Methods in Molecular Biology, vol. 1738, https://doi.org/10.1007/978-1-4939-7643-0_9, © Springer Science+Business Media, LLC, part of Springer Nature 2018

not be viewed as a chemometric technique but as a physiological analysis leading to biologically relevant conclusions. As such, it is a multistep procedure, and its standardization requires the integrated optimization of pre-analytical, analytical, and computational stages [4]. The pre-analytical steps include the experimental design that needs to take into consideration both the biological and the analytical constraints of the study and of the involved omic analysis/es and the sample collection, handling, quenching, and storage protocols. While special care should be given to the sample collection and handling protocol in any omic analysis to avoid major perturbations of the samples physiological state, the optimization of this step becomes even more important in metabolomics, taking into consideration that metabolism is a very dynamic cellular process. Interestingly, in metabolomics, quenching and metabolite extraction, the first analytical step, could take place simultaneously in the case of polar metabolites, as methanol used for quenching is an extraction agent of polar metabolites too. In most cases, for consistency between extraction protocols and different omic analyses, it is preferable that the quenching is separated from the extraction step. Especially for tissues, quenching is carried out by fast freezing of the samples in liquid nitrogen and stored at −80 °C until further analysis. If transfer is required, the quenched samples should be transported on dry ice (*see* **Note 1**).

There is no extraction reagent for the entire metabolome; methanol- and chloroform-based protocols are mainly used for the extraction of polar and nonpolar metabolites, respectively, and optimized protocols have been proposed for a variety of systems [2]. Concerning the analytical techniques for the acquisition of the metabolic profile, in the case of untargeted metabolomic analyses, mass spectrometry (MS) in conjunction with gas (GC) or liquid chromatography (LC) is sometimes preferred over nuclear magnetic resonance (NMR) spectroscopy, because they are more sensitive and high throughput [2, 5]. However, mass spectrometry-based metabolic profiling contain a large number of unidentified peaks, which cannot be related to the metabolic physiology of the investigated system [2, 5]; thus, educated methods for peak identification are required. Between GC-MS and LC-MS, the advantages of the former include the higher chromatographic resolution in the gas compared to the liquid phase and the larger databases of identified peaks, because of its longer use in clinical chemistry practice. Moreover, GC-MS can measure diverse classes of compounds with one type of column, and the electron ionization mode involves fragmentation of the molecules according to their structure, which contributes to their identification and lessens the need for tandem mass spectrometry. For all these reasons, GC-MS is an integral equipment of a mass spectrometry metabolomics facility. It has been widely used for untargeted metabolomic analyses of the primary metabolism activity, as it can quantify compounds of molecular weight smaller than 600 a.u.

However, GC-MS metabolomics imposes an additional analytical step, which concerns the derivatization of the extracted metabolites into volatile and thermally stable derivatives [2]. This is a crucial procedure that needs to be carefully standardized, especially in untargeted metabolomics, and affects both the analytical and the computational protocols of the analysis [2, 6]. This is due to the fact that the derivatization kinetics is different between the compounds and depends on the composition of each metabolite extract (matrix effects) [2, 6]. Thus, the metabolic profile acquisition of all derivatized samples in an experiment should be carried out at or after the derivatization time at which all original metabolites in any sample have been fully transformed into a derivative. This time may differ between samples and should be identified through specific experiments [2, 6]. Moreover, there exists no derivatization method that leads to one derivative per metabolite for all compound classes [2, 6]. Some metabolites may form multiple derivatives, the relative concentration of which changes with the duration of the derivatization [2, 6]. The most widely used derivatization method in untargeted GC-MS metabolomics, as it covers a wide range of metabolite classes, is a two-step procedure, involving (a) the methoximation of the ketone group-containing metabolites into two stable methoximes of constant concentration ratio, and (b) the silylation of all metabolites, including the methoximes formed in the previous step, into their trimethylsilyl (TMS) derivatives [2, 6]. In this case, the amine group-containing metabolites can form multiple derivatives produced serially as the derivatization progresses [2, 6]. The constant ratio between the two methoxime derivatives of ketone group-containing metabolites between samples could be used as a quality control criterion that all samples have been run under the same derivatization and equipment acquisition conditions. It is underlined that it is the only such criterion that can be applied post-experimentally, not requiring quality control samples [2, 6]. Moreover, estimating the effective derivative peak area of each amine group containing metabolite that is directly proportional to the concentration of the original metabolite in the samples requires specialized normalization methods [2, 6] that need to be appropriately used in the computational part of the protocol [7]. However, the application of these normalization methods requires the availability of certain measurements; thus, the experimental protocol should be appropriately modified so that the profiles of certain samples could be quantified at multiple derivatization times [2, 6]. In this way, the metabolomic analysis presented here is metabolite-centric and not feature- or derivative-centric.

Taking all these special characteristics of untargeted GC-MS metabolomics into consideration, we present a standardized protocol that can be applied to any biological system, with certain adaptations depending on the type of the sample. Such adaptations are indicated either in the protocol and/or the accompanying notes. The protocol integrates the experimental and the computational parts, with the former including the polar metabolite extraction, the derivatization, and the GC-MS metabolic profile acquisition and then the metabolite derivative identification/quantification, data normalization and filtering, the multivariate statistical analysis and the metabolic network reconstruction. Our research group has already used this protocol to acquire the metabolic profiles of plant [8–10] and animal [11, 12] tissues, cell cultures [13, 14], and blood plasma samples [5, 15].

2   Materials

2.1   Consumables

1. Tips for automatic pipettes.

2. Glass tips.

3. 2 mL microcentrifuge tubes (Eppendorf type)—for low-volume samples.

4. Pellet micropestles (for the homogenization of low-volume samples in Eppendorf-type tubes).

5. 15 mL centrifuge tubes (Falcon type)—for high-volume samples.

6. 1.5 mL high-recovery autosampler glass vials with PTFE crimp caps.

2.2   Reagents

1. HPLC grade water.

2. HPLC grade methanol.

3. Adonitol/ribitol (as internal standard).

4. U-^{13}C-D-Glucose (as internal standard).

5. Pyridine anhydrous.

6. Methoxylamine hydrochloride (98+%).

7. N-methyl-N-(trimethylsilyl)trifluoroacetamide   (MSTFA) (97+%).

2.3   Wet Lab Equipment

1. Glass homogenizer with Teflon pestle (Thomas Scientific, Swedesboro, NJ, USA).

2. Automatic pipettes.

3. Pipettes with glass tips (Drummond Scientific, Broomall, PA, USA).

4. Water bath.

5. Cooling centrifuge (Thermo Fisher Scientific, Waltham, MA, USA).

6. Vacuum centrifuge/concentrator (Thermo Fisher Scientific, Waltham, MA, USA).

7. Benchtop shaker incubator (New Brunswick Scientific, Einfield, CT, USA).

2.4 GC-MS Equipment and Consumables

1. 3800 series GC (Varian, Palo Alto, CA, USA—now Bruker, Billerica, MA, USA).

2. Saturn 2200 series MS ion trap (Varian, Palo Alto, CA, USA—now Agilent, Santa Clara, CA, USA)—the protocol provided can be applied to any GC-MS equipment.

3. CP-8410 autosampler (Bruker, Billerica, MA, USA).

4. GC capillary column: Zebron, ZB-50, 30 m × 0.25 mm ID × 0.25 μm (Phenomenex, Torrance, CA, USA).

5. Helium 99.999% carrier gas (Air Liquide Hellas, Athens, Greece).

2.5 Software

1. MS workstation v. 6.5 (Varian, Palo Alto, CA, USA—now Agilent, Santa Clara, CA, USA)—any other software for peak identification compatible with the available GC-MS equipment could be used.

2. The National Institute of Standards and Technology (NIST) mass spectral search program for the NIST/EPA/NIH mass spectral library.

3. Microsoft Excel 2010 (Microsoft, Redmond, WA, USA).

4. M-IOLITE GC-MS metabolomic data repository, normalization, and unknown peak identification software platform (v1 beta)—available upon request for academic users at *miolite. iceht.forth.gr* [7].

5. TM4/MeV omic data analysis open-source software, v.4.9.0 [16, 17].

3 Methods

3.1 Analytical Protocol

Intact or Lyophilized Tissues or Cell Pellets

3.1.1 Extraction of Polar Metabolites from

1. (*In the case of tissue or cell pellets available in large quantity, if this step is not carried out before quenching*)

Weigh the intact tissue [11, 12] or the selected amount of lyophilized tissue [8–10] or cell pellet (if a large quantity is available, see industrial cell culture application [13]) rapidly before it thaws; the weight is required for the addition of the appropriate amount of methanol/water and internal standards below.

(*In the case of cell pellet from a certain culture volume, e.g., from one petri dish culture—small quantity available*)

The amount of cells in each sample is estimated from the quantified protein content (*need for the application of a protein content quantification method*)—if all samples originate from the same culture volume, it is usually opted to add the same amount of internal standards and methanol/water in excess to all and then normalize the profiles based on the respective protein content at the normalization stage (*see* **Note 2**).

2. Transfer the tissue or cell pellet (available in large quantity) sample to a glass homogenizer, and add ice-cold HPLC grade methanol in a ratio of 22 mL per g of sample (e.g., add 2.2 mL of cold methanol to 100 mg of tissue) [8, 9].
   (*In the case of cell pellet from a certain culture volume, e.g., from one petri dish culture—small quantity available*)
   Add 0.5–1 ml methanol to the cell pellet from one petri dish of mammalian cell cultures (protein content ~5 mg protein) directly into the Eppendorf-type tube (as transfer in these cases is not preferable), and use a pellet micropestle for its homogenization (**step 5** below).

3. Add internal standards in a ratio of 0.1 µg adonitol/ribitol and 0.2 µg of U-^{13}C-D-glucose per 1 mg of tissue/cell pellet/protein content.

4. Homogenize the sample thoroughly.

5. In the case that the glass homogenizer has been used, transfer the homogenate to a 2 mL Eppendorf- or a 15 mL Falcon-type tube (depending on the sample volume).

6. Incubate in water bath at 70 °C for 15–20 min.

7. Add HPLC grade water in equal volume to the methanol initially added in **step 2**, and mix gently, turning the tubes upside down to enable the mixing of both phases.

8. Centrifuge the samples at 10,000 × g for 10 min at 4 °C.

9. Transfer the supernatant of each sample into a clean 2 mL Eppendorf- or a 15 mL Falcon-type tube (depending on the supernatant volume).

10. Repeat centrifugation as many times as required until no precipitate is visible.

11. Transfer all the supernatant into a high-recovery autosampler glass vial (if the supernatant volume is small), or, if possible, divide it into 2–4 replicates.

12. Vacuum dry the supernatant samples.

13. Cap each vial with a PTFE crimp cap and store at 4 °C until further analysis.

Blood Plasma or Other Liquid (e.g., Culture Medium) Samples

1. Add 200 µL of blood plasma into a 2 mL Eppendorf tube containing 0.6 mL of ice-cold methanol and 2 µg of adonitol/ribitol and 4 µg of U-^{13}C-D-glucose as internal standards (*see* **Note 3**).

2. Mix the sample.

3. If transfer is required, the samples could be transported in ice as proteins have been denatured by methanol.

4. Follow **steps 6–13** as described previously in the section for intact or lyophylized tissues or cell pellets.

3.1.2 TMS Derivatization

1. Vacuum dry each stored at 4 °C vial for 30 min to remove any remaining humidity (*see* **Note 4**).

2. Add 50 µL of 20 mg/mL solution of methoxylamine hydrochloride in pyridine (*see* **Notes 5** and **6**) using glass pipette tips, and mix gently (*see* **Notes 4** and **6**); the volume of the solution to be added may vary from 30 µL for very small amounts of dry extracts to 150 µL for larger samples—preliminary experiments should be carried out to determine the optimal volume for a particular sample type balancing between the need for in excess availability of the derivatization reagents and avoiding any large dilution of the dry extract [2, 6].

3. Incubate in a shaker incubator at 40 °C for 90 min.

4. Add MSTFA at twice the volume of the methoxylamine hydrochloride solution added in **step 2** using glass pipette tips and mix gently.

5. Incubate at 40 °C for at least 6 h. This is the derivatization time at which all metabolites in a sample are estimated to have been fully transformed into at least one of their TMS derivatives [2, 6]. This time may vary between biological systems and could be optimized through a preliminary experiment at which the metabolic profile of a certain system is measured multiple times from 15 min to 10 h of derivatization duration [2]. However, based on our group's experience with various systems, 6 h of derivatization could be considered a "universal" time for most biological sample types for the standardization of this step of the protocol.

6. Place the vial on the autosampler for the sample to be used for metabolic profile acquisition.

   Significant note: In an optimized GC-MS metabolic profile acquisition protocol, taking into consideration the constraint of **step 5**, four [4] samples can be quantified per day with three repetitions per sample and three runs of the solvent (pyridine) to clean the column between samples (see the acquisition section below). In this optimized protocol, the incubation time of **step 5** is equal to 9 h.

3.1.3 GC-MS Metabolic Profile Acquisition

1. Set the injector temperature to 230 °C and the detector transfer line, trap, and manifold temperatures to 250 °C, 220 °C, and 70 °C, respectively. These parameters can be appropriately adapted in a GC-quadrupole MS.

2. Set the helium carrier gas flow rate to 1 mL/min.

3. Program the GC oven temperature gradient as follows: set initial oven temperature at 70 °C, hold for 5 min, and then increase temperature up to 310 °C at a rate of 5 °C/min. Finally, hold for 3 min. The total run time is 56 min.

4. Set the split ratio to the value optimized according to the expected metabolite concentration range in the samples under investigation. A split ratio between 1:40 and 1:25 is suitable for most sample types; it requires optimization through preliminary tests.

5. Operate the electron ionization source at −70 eV.

6. Operate the mass spectrometer in scan mode over a mass range of 50–600 m/z; start mass spectrum acquisition 4 min after the initialization of the run to allow for the solvent to be eluted.

7. Inject 1 µL from a sample to acquire its metabolic profile (*see* **Note** 7).

8. Acquire the metabolic profile of each sample at least twice in two consecutive runs. At least two runs of solvent (pyridine) should be carried out between biological samples to clean the column. Figure 1 shows the profiles of a baby hamster kidney (BHK) cell culture [14], C57BL/6J male mouse cortex, and tomato leaf sample at 11 h of derivatization.

9. The metabolic profile of at least one sample per experimental batch (optimally one per each week of runs) should be quantified at least five times at derivatization durations from 9 h to 15 h. These measurements are required for the application of specialized normalization methods on the profiles of the amine group-containing metabolites (*see* below).

3.2 Computational Protocol

3.2.1 Peak Identification and Quantification

1. Collect the files storing the MS-reconstructed chromatograms of the metabolic profiles of the various samples, and proceed to peak identification and quantification using a relevant software, usually associated with the available GC-MS equipment.

Special note : A "metabolite-centric" compared to a "feature-centric" approach in the analysis of the MS-reconstructed chromatograms is recommended [2, 18]. It is more biologically relevant, enabling the identification and exclusion of artifacts at the normalization and filtering step (*see* below). In the "feature-centric" approach, one identifies and quantifies every ion peak in the MS-reconstructed chromatogram, while in the "metabolite-centric" approach, the profile is scanned based on a standardized database of peaks, and each metabolite derivative is characterized and quantified based on one of its ion peaks, called "marker or quantifying ion." One can use feature-scanning software to automatically scan the profile and then filter the data based on the metabolite database or scan the profile only for the peaks in the

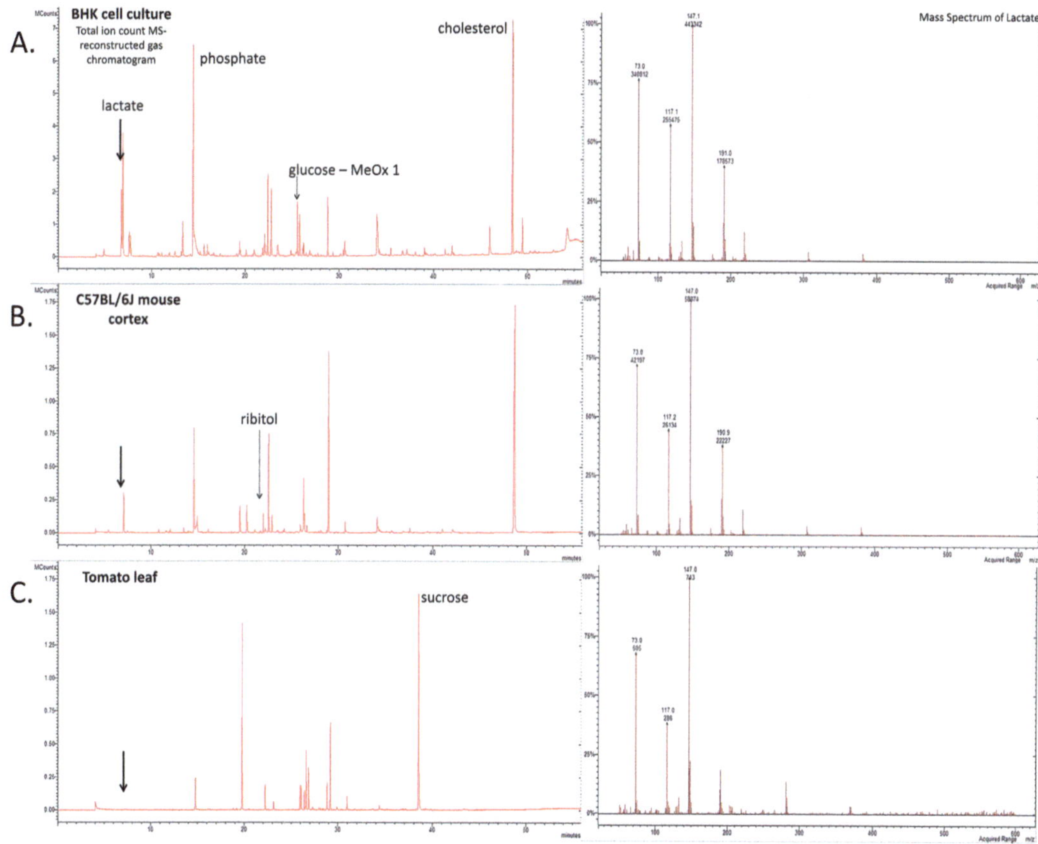

Fig. 1 Total ion count MS-reconstructed gas chromatograms for polar metabolite extracts (left panel) and the respective mass spectrum of the lactate 2TMS derivative (right panel) of samples from (**a**) industrial-scale *BHK* cell culture [14], (**b**) C57BL/6 J male mouse cortex, and (**c**) tomato leaf. The straight arrow shows the peak of the lactate 2TMS derivative in each chromatogram. The metabolites corresponding to some major peaks are also denoted. In all three samples, the internal standard ribitol has been added in the same relative concentration with respect to the sample weight (see text)

database based on an accordingly developed method. Involvement of an expert user understanding the biochemistry of the investigated biological system and the specifics of the GC-MS analysis is required at this stage to appropriately guide and supervise the peak identification and quantification process.

2. Merge all profiles into one file that can be easily presented and computationally processed. It is advisable that the data are submitted to standardized repositories to enable meta-analysis studies. *MetaboLights* (http://www.ebi.ac.uk/metabolights/) emerges as the reference repository for metabolomic data.

3.2.2 Data Normalization and Filtering

1. Investigate whether all profiles were acquired at the same analytical process conditions by estimating the ratio of the two methoxime peaks of [U-^{13}C]—glucose used as internal standard. If available, the ratio of the two methoxime peaks of

other metabolites could also be estimated. If these ratios remain constant among the samples, subject only to random variance, then constant conditions can be validated throughout the experimental process, and the profiles are directly comparable [2, 6]. If large deviation is indicated in any profiles, then these are excluded from further analysis. This quality control criterion does not require the availability of any quality control (QC) samples and can be used post-experimentally to evaluate the comparability of profiles that have been acquired at different experimental batches.

2. Normalize the marker ion peak area of the detected compounds with the marker ion peak area of the internal standard. Ribitol has been proven a reliable internal standard for most biological systems and applications, as it is not endogenously produced or in very small quantities that do not affect the quantification of the exogenously provided standard [2, 6, 8–15]; the latter statement requires verification in every new system under investigation. Among the two marker ions of ribitol (217 or 319), the latter is usually preferred at this normalization step as its smaller peak area does not bias the estimation of the relative peak areas (RPAs) of the least abundant metabolites toward very small values that can "skew" the subsequent data analysis.

3. Estimate the cumulative (effective) peak area of the known amine group-containing metabolites as the weighted sum of their derivative peak areas [2, 6]. The weights are estimated from the profiles of the samples that were quantified at least five times within 9 h and 15 h of derivatization (*see* **step 9** in GC-MS metabolic profile acquisition), based on the algorithm described in [2, 6]; *see* Fig. 2.

4. Filter out of the dataset technical and mathematical artifacts to avoid skewing the subsequent data analysis. Specifically, exclude:

   (a) Peaks with small signal-to-noise ratio

   (b) Peaks with significant carryover

   (c) Technical artifacts from column bleeding and/or reagents (*see* **Note 7**)

   (d) Peaks that are inconsistently detected among samples (with large number of missing values)

   (e) Peaks with mean variation among technical replicates in all samples that is greater than 30%

A

Equation for the estimation of the weights to be used in the calculation of the effective relative peak area (RPA) of the amine-group containing metabolites in each sample

$$\begin{bmatrix} w_{deriv1} \\ w_{deriv2} \\ w_{deriv3} \end{bmatrix} = \left(\begin{bmatrix} RPA_{deriv1}{}^{t_1} & RPA_{deriv2}{}^{t_1} & RPA_{deriv3}{}^{t_1} \\ RPA_{deriv1}{}^{t_2} & RPA_{deriv2}{}^{t_2} & RPA_{deriv3}{}^{t_2} \\ RPA_{deriv1}{}^{t_3} & RPA_{deriv2}{}^{t_3} & RPA_{deriv3}{}^{t_3} \\ RPA_{deriv1}{}^{t_4} & RPA_{deriv2}{}^{t_4} & RPA_{deriv3}{}^{t_4} \\ RPA_{deriv1}{}^{t_5} & RPA_{deriv2}{}^{t_5} & RPA_{deriv3}{}^{t_5} \end{bmatrix} \right)^{-1} \times \begin{bmatrix} b \\ b \\ b \\ b \\ b \end{bmatrix}$$

$$RPA_{effective} = \sum_i w_i \cdot RPA_{deriv_i}$$

B

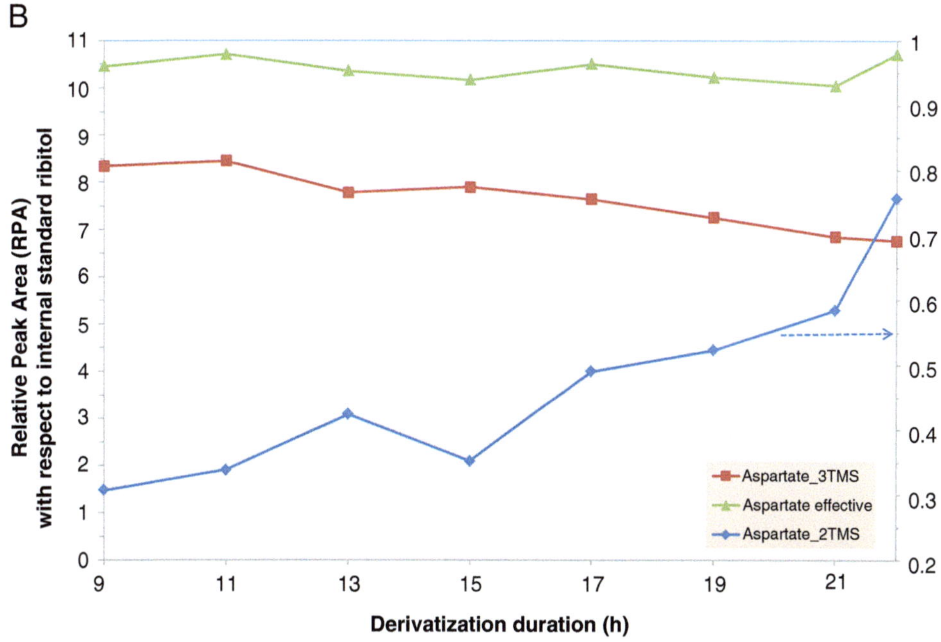

Fig. 2 (**a**) The algorithm for the estimation of the weights for the derivative peak areas of an amine group containing metabolite required for the estimation of its effective peak area in each sample (see equation at the bottom) [2, 6]; (**b**) the profile of the relative peak areas (RPAs) of the two derivatives and of the estimated effective RPA of aspartate over ribitol (internal standard) in a Balbc/J male mouse cortex sample over eight [8] different derivative durations. In (**a**), the algorithm is shown for a metabolite with three [3] derivatives (deriv$_i$) and a normalization sample, the profile of which was acquired at five different derivatization durations (t$_i$)—the algorithm should be appropriately adjusted for different conditions. The "b" constant is defined in the same order of magnitude either of the largest derivative or the sum of derivatives (if different) in the investigated profiles, as explained in [2, 6]

(f) Peaks corresponding to unknown amine group containing metabolites, as it is not possible to estimate their effective peak area—see point 3 above—and are subject to derivatization biases [2, 6]

(g) The peak of the least abundant methoxime derivative of any ketone group containing metabolite of known identity, if both methoximes have been quantified; as the abundances of the two methoximes of a ketone group containing metabolite are linearly dependent, inclusion of both peak areas in the final dataset will introduce bias in data analysis [2, 6].

After this normalization and filtering step, the final dataset concerns peak areas of metabolites and not metabolite derivatives (the only relevant deviation may be introduced from the presence of unknown ketone group containing metabolites for which both methoxime peaks have been identified; however, such pairs are usually detectable despite unknown identity, and filtering of one of the two occurs in step g above). Our group has recently developed a software platform, called M-IOLITE [7], including a module for GC-MS metabolomic data normalization and filtering as previously described (available upon request for academic users at *miolite.iceht.forth.gr*).

3.2.3 Multivariate Statistical Analysis and Metabolic Network Reconstruction

1. The final metabolite profile dataset after normalization and filtering can be used for further analysis to extract biologically relevant conclusions. As for other omic profiles, multivariate statistical analysis methods can be applied to identify correlations between metabolites and/or physiological conditions. Specifically, supervised or unsupervised clustering algorithms can group metabolites and/or samples of similar profiles based on a variety of profile distance metrics (e.g., hierarchical clustering analysis (HCL) [19]). Moreover, methods like principal component analysis (PCA) [20] or partial least squares (PLS) regression [21] can be used to lower the dimensionality of the problem, enabling the visualization of the differences between the profiles in a 3-D space, providing also information about the metabolites that contribute significantly to these differences. The metabolites with significantly differential concentration between sets of metabolic profiles can be identified with the multivariate significance analysis algorithm, called significance analysis of microarrays (SAM) [22]. SAM has been appropriately tuned for omic data, since it does not require that they follow a particular distribution (as the t-test or F-test) and estimated the false discovery rate for each threshold of significance.

   There exist many software platforms, publicly or commercially available, which incorporate many of these bioinformatics and data mining algorithms, some built around an interface that is friendly for a nonexpert in computational analysis. The open-source software TM4 Multiple Experiment Viewer (MeV) (http://mev.tm4.org) provides a user-friendly platform with a big gallery of multivariate statistics algorithms. MATLAB (https://www.mathworks.com/products/matlab/) has also developed an extensive bioinformatics toolbox, which combines the functionality of the MATLAB environment and language. Many developers and computational users opt for the use of relevant modules in R language (*R* Project for Statistical Computing; https://www.r-project.org/), most biologically relevant having been incorporated in the Bioconductor open-software suite (https://www.bioconductor.org/).

2. A major step in metabolomic data analysis is the interpretation of the results in the context of metabolic network structure and regulation. Fortunately, compared to biomolecular networks at the transcriptional and translational levels, there exists extensive knowledge about the structure and function of metabolic pathways. This information enables the estimation of the relative change between pathway fluxes even in the cases that not all metabolites in the pathways are quantified. Information for the reconstruction of the metabolic pathways for different organisms or tissues can be obtained from metabolic databases, e.g., KEGG (Kyoto Encyclopedia of Genes and Genomes)—GenomeNet (http://www.genome.jp/kegg/); ExPASy: SIB Bioinformatics Resource Portal (https://www.expasy.org/); and BioCyc Database Collection (http://biocyc.org/), or metabolomic databases, e.g., HMDB (Human Metabolome Database) (http://www.hmdb.ca/). Moreover, there exist genome-scale reconstructed metabolic networks for various organisms, such as Recon 2 for the human (http://www.ebi.ac.uk/biomodels-main/MODEL1603150001), the mouse (http://www.ebi.ac.uk/compneur-srv/biomodels-main/MODEL1507180055), and various tissues, such as the brain [2, 11, 12].

4 Notes

1. Samples can be stored at −80 °C for at least 6 months; actually the maximum storage duration that excludes metabolite degradation and changes in the acquired metabolic profiles is currently under investigation, especially in connection with biobanking. Avoid any sample thawing before the samples are processed for extraction. It would be preferable if the samples are extracted and the metabolite extracts can be stored for longer times at 4 °C.

2. For fibrous samples, like plant leaves, grinding the sample to powder with mortar, pestle, and liquid nitrogen prior to extraction is advised.

3. EDTA (and not heparin) should be used as anticoagulant for blood sample acquisition; it is considered to affect the composition of the sample less and its peak can be isolated from the rest of the metabolic profile without affecting the measurements of other metabolites. Citrate vacutainers should not be used.

4. MSTFA is not active for derivatization in the presence of water. Please make sure that the metabolite extracts are appropriately dried and MSTFA has been appropriately stored. Moreover, all reagents used for the derivatization of the extracts should be anhydrous.

5. Pyridine and MSTFA are volatile and hazardous. They should be handled quickly in a fume hood applying proper precautions.

6. Pyridine reacts with plastic. Avoid any use of plastic consumables for handling and storing the samples in all protocol steps after the initialization of pyridine use. The caps of the autosampler vials should contain only PTFE and not silicone, as the latter can react with pyridine and contaminate the samples.

7. Before each experiment, quantify the profile of a negative control sample to identify the peaks and their areas that originate from the reagents and the consumables used in the process (including column bleeding), so that they can be excluded from the metabolic profile and further analysis.

References

1. Fiehn O (2002) Metabolomics–the link between genotypes and phenotypes. Plant Mol Biol 48(1–2):155–171. https://doi.org/10.1007/978-94-010-0448-0_11

2. Kanani H, Chrysanthopoulos PK, Klapa MI (2008) Standardizing GC-MS metabolomics. J Chromatogr B 871(2):191–201. https://doi.org/10.1016/j.jchromb.2008.04.049

3. Patti GJ, Yanes O, Siuzdak G (2012) Innovation: metabolomics: the apogee of the omics trilogy. Nat Rev Mol Cell Biol 13(4):263–269. https://doi.org/10.1038/nrm3314

4. Vasilopoulou CG, Margarity M, Klapa MI (2016) Metabolomic analysis in brain research: opportunities and challenges. Front Physiol 7:183. https://doi.org/10.3389/fphys.2016.00183

5. Spagou K, Theodoridis G, Wilson I et al (2011) A GC-MS metabolic profiling study of plasma samples from mice on low- and high-fat diets. J Chromatogr B 879(17–18):1467–1475. https://doi.org/10.1016/j.jchromb.2011.01.028

6. Kanani HH, Klapa MI (2007) Data correction strategy for metabolomics analysis using gas chromatography-mass spectrometry. Metab Eng 9(1):39–51. https://doi.org/10.1016/j.ymben.2006.08.001

7. Maga-Nteve C, Klapa MI (2016) Streamlining GC-MS metabolomic analysis using the M-IOLITE software suite. IFAC-PapersOnLine 49(26):286–288. https://doi.org/10.1016/j.ifacol.2016.12.140

8. Dutta B, Kanani H, Quackenbush J, Klapa MI (2009) Time-series integrated "omic" analyses to elucidate short-term stress-induced responses in plant liquid cultures. Biotechnol Bioeng 102(1):264–279. https://doi.org/10.1002/Bit.22036

9. Kanani H, Dutta B, Klapa MI (2010) Individual vs. combinatorial effect of elevated CO_2 conditions and salinity stress on *Arabidopsis thaliana* liquid cultures: comparing the early molecular response using time-series transcriptomic and metabolomic analyses. BMC Syst Biol 4:177. https://doi.org/10.1186/1752-0509-4-177

10. Tooulakou G, Giannopoulos A, Nikolopoulos D et al (2016) "Alarm photosynthesis": calcium oxalate crystals as an internal CO_2 source in plants. Plant Physiol 171(4):2577–2585. https://doi.org/10.1104/pp.16.00111

11. Constantinou C, Chrysanthopoulos PK, Margarity M, Klapa MI (2011) GC-MS metabolomic analysis reveals significant alterations in cerebellar metabolic physiology in a mouse model of adult onset hypothyroidism. J Proteome Res 10(2):869–879. https://doi.org/10.1021/pr100699m

12. Maga-Nteve C, Vasilopoulou CG, Constantinou C et al (2017) Sex-comparative study of mouse cerebellum physiology under adult-onset hypothyroidism: the significance of GC-MS metabolomic data normalization in meta-analysis. J Chromatogr B 1041-1042:158–166. https://doi.org/10.1016/j.jchromb.2016.12.016

13. Chrysanthopoulos PK, Goudar CT, Klapa MI (2010) Metabolomics for high-resolution monitoring of the cellular physiological state in cell culture engineering. Metab Eng 12(3):212–222. https://doi.org/10.1016/j.ymben.2009.11.001

14. Vernardis SI, Goudar CT, Klapa MI (2013) Metabolic profiling reveals that time related physiological changes in mammalian cell perfusion cultures are bioreactor scale independent. Metab Eng 19:1–9. https://doi.org/10.1016/j.ymben.2013.04.005

15. Gkourogianni A, Kosteria I, Telonis AG et al (2014) Plasma metabolomic profiling suggests early indications for predisposition to latent insulin resistance in children conceived by ICSI. PLoS One 9(4):e94001. https://doi.org/10.1371/journal.pone.0094001

16. Saeed AI, Bhagabati NK, Braisted JC et al (2006) TM4 microarray software suite. Methods Enzymol 411:134–193. https://doi.org/10.1016/S0076-6879(06)11009-5

17. Saeed AI, Sharov V, White J et al (2003) TM4: a free, open-source system for microarray data management and analysis. BioTechniques 34(2):374–378

18. Allwood JW, Erban A, de Koning S et al (2009) Inter-laboratory reproducibility of fast gas chromatography-electron impact-time of flight mass spectrometry (GC-EI-TOF/MS) based plant metabolomics. Metabolomics 5(4):479–496. https://doi.org/10.1007/s11306-009-0169-z

19. Eisen MB, Spellman PT, Brown PO, Botstein D (1998) Cluster analysis and display of genome-wide expression patterns. Proc Natl Acad Sci U S A 95(25):14863–14868. https://doi.org/10.1073/pnas.95.25.14863

20. Raychaudhuri S, Stuart JM, Altman RB (2000) Principal components analysis to summarize microarray experiments: application to sporulation time series. Pac Symp Biocomput 2000:455–466

21. Maitra S, Yan J (2008) Principle component analysis and partial least squares: two dimension reduction techniques for regression. In: 2008 Casualty actuarial society discussion paper program–applying multivariate statistical models. Casualty Actuarial Society, Quebec, pp 79–90

22. Tusher VG, Tibshirani R, Chu G (2001) Significance analysis of microarrays applied to the ionizing radiation response. Proc Natl Acad Sci U S A 98(9):5116–5121. https://doi.org/10.1073/pnas.091062498

Chapter 10

Rat Fecal Metabolomics-Based Analysis

Olga Deda, Helen G. Gika, and Georgios A. Theodoridis

Abstract

Fecal metabolomics-based analysis indisputably constitutes a very useful tool for elucidating the biochemistry of digestion and absorption of the gastrointestinal system. Fecal samples represent the most suitable, non-invasive, specimen for the study of the symbiotic relationship between the host and the intestinal microbiota.

It is well established that the balance of the intestinal microbiota changes in response to some stimuli, physiological such as gender, age, diet, exercise and pathological such as gastrointestinal and hepatic disease. Fecal samples have been analyzed using the most widespread analytical techniques, namely, NMR spectroscopy, GC-MS, and LC-MS/MS. Rat fecal sample is a frequently used and particularly useful substrate for metabolomics-based studies in related fields. The complexity and diversity of the nature of fecal samples require careful and skillful handling for the effective quantitative extraction of the metabolites while avoiding their deterioration. Parameters such as the fecal sample weight to extraction solvent volume, the nature and the pH value of the extraction solvent, and the homogenization process are some important factors for the optimal extraction of samples, in order to obtain high-quality metabolic fingerprints, using either untargeted or targeted metabolomics.

Key words Metabolomics, Sample preparation, Fecal samples, Rats, NMR, GC-MS, LC-MS/MS, Fecal extract

1 Introduction

Fecal sample is a useful bio-specimen in chemical/biochemical point of view. Fecal sample analysis is a powerful tool in prognosis and diagnosis for a plethora of gastrointestinal diseases [1, 2]. Raw fecal samples require appropriate preparation in order to produce reliable conclusions [3].

Sample preparation is a very critical aspect in metabolomics-based analysis and can have significant effects on the quality of the obtained results [4–6]. The peculiarities of the fecal sample in contrast with common samples include complexity, heterogeneity, and a high concentration of nondigested macromolecules [3, 7, 8].

Georgios A. Theodoridis et al. (eds.), *Metabolic Profiling: Methods and Protocols*, Methods in Molecular Biology, vol. 1738, https://doi.org/10.1007/978-1-4939-7643-0_10, © Springer Science+Business Media, LLC, part of Springer Nature 2018

The composition of the fecal sample is largely influenced by dietary factors and could reflect the nutrient's metabolism by gut microbiota [9]. Metabolomics-based analysis provides a detailed picture of the fecal sample metabolome, which is derived from co-metabolism of gut microbiota and the host organism. Fecal samples should undergo special treatment, due to their nature, for the avoidance of negative effects on the analytical systems.

Fecal sample preparation is a determining factor for the quality of the analysis, regardless of the analytical technique subsequently applied. Inappropriate sample preparation could lead to ineffective, non-repeatable, and poor extraction, thus affecting the obtained metabolic profile. The golden rule is a fast and repeatable, generic (nonselective) sample preparation process, capable of extracting metabolites of different chemical classes. Each step of this process is of great importance.

Undoubtedly, the ratio of fecal sample weight to extract solvent volume has the greatest impact [8, 10, 11]. However, this does not necessarily mean that denser samples lead to better results, because there is a possibility that high density will create interferences in the analytical system.

Solvent extraction buffer is also a very strong factor affecting the fecal metabolic profiling obtained by the most commonly used analytical techniques, namely, nuclear magnetic resonance (NMR) spectroscopy, gas chromatography-mass spectrometry (GC-MS), and liquid chromatography-mass spectrometry (LC-MS/MS). The nature of the extraction solvent and the pH value of the extraction solvent strengthen or weaken the extraction of particular metabolites. For example, inappropriate choice of extraction solvent solution pH value may lead to the deterioration of fecal extracts through hydrolysis of some metabolites and affects the extraction capability of ionizable metabolites.

Furthermore, homogenization, filtration, and centrifugation are also parameters which can prove critical [12]. Homogenization by sonication or mechanical smashing, by using a mechanical crusher, or TissueLyser leads to better results, especially in NMR-based metabolomics [11, 12]. Disaggregating fecal sample prior to the homogenization step could also enhance extraction efficacy [8].

Although filtration is not as important a parameter as the previously mentioned processes, with regard to affecting the fecal metabolic profiling, it is a proper (if not necessary) practice for removing particulates thus protecting the analytical system, especially in LC-MS/MS-based metabolic profiling [13]. In order to remove particulates, usually, two cycles of centrifugation are required: the first in the early steps of the process and the second immediately before the subsequent analysis.

Another practice, rather controversial, in fecal sample preparation is lyophilization. Although lyophilization eliminates the water content [14], its use is not recommended as it has been proven to affect the obtained fecal metabolic profiling. For example, it could lead to reduced ammounts of short-chain fatty acids (SCFAs) and potentially of other volatile compounds [12, 15], thus editing the profile especially in GC-MS mode.

In addition, more than one extraction cycle could be performed, and the combination of extracts could take place [11]. This practice is not recommended for the sake of simplicity of the sample preparation process.

Finally, depending on the technique by which fecal samples are analyzed, other conditions and parameters may affect the analytical performance. For example, in NMR spectroscopy-based metabolomics analysis, PBS buffer preparation (preparation in D_2O, pH value, and salt addition) appreciably reflects the quality of the collected spectra. The pH value of PBS may significantly affect the behavior of the analytes and the chemical shift reference used.

Undoubtedly, for GC-MS based analysis the derivatization process of the fecal sample is a major determining factor. Critical factors such as residual moisture, insufficient derivatization reagents, and shorter or longer incubation, at lower or higher temperature, greatly affect the obtained chromatograms [16, 17].

Selection of an extraction solvent, similar to the mobile phase, could result in better peaks in liquid chromatography, while very concentrated fecal extract could block the column and prevent further analysis.

2　Materials

2.1　Reagents

1. 1-propanol (LC-MS grade).
2. Acetonitrile (LC-MS grade).
3. Methanol (LC-MS grade).
4. Methanol (HPLC grade) (*see* **Note 1**).
5. Deuterium oxide (D_2O) 99.96%.
6. Sodium dihydrogen phosphate monohydrate ($H_2O \cdot NaH_2PO_4$).
7. Disodium hydrogen phosphate anhydrous (Na_2HPO_4) (*see* **Note 2**).
8. Sodium chloride, granular/USP/FCC (*see* **Note 3**).
9. Pyridine anhydrous 99.8%.
10. Methoxyamine hydrochloride (MeOX).
11. N-methyl-N-(trimethylsilyl)trifluoroacetamide (MSTFA).
12. Trimethylchlorosilane (TMCS).
13. Trimethylsilylpropanoic acid (TSP) (sodium salt of deuterated molecule).

14. Ammonium formate, for MS ≥99.0% (*see* **Note 4**).

15. Formic acid eluent additive for LC-MS/MS (*see* **Note 5**).

16. H_2O (LC-MS/MS grade).

2.2 Laboratory Equipment

1. −80 °C freezer.

2. Millipore purification system for water.

3. Electronic analytical balance.

4. Mini shaker laboratory vortex.

5. TissueLyser.

6. Ultrasonic tissue processor.

7. Micro refrigerated centrifuge.

8. Pure nitrogen gas generator.

9. Adjustable variable volume pipettes of 1000, 100, 20, and 10 mL and pipette tips.

10. Eppendorf tubes of 2 mL.

11. Screw cap glass vials of 1.5 mL.

12. Syringes of 2.5 mL.

13. 25 mm diameter sterile syringe filters PTFE with 0.22 μm pore size.

14. NMR sample tubes, 5 mm, 7″ length (or appropriate size for the spectrometer probe used).

15. GC-MS vials, caps, and insert vials 250 μL.

16. LC-MS/MS glass vials, caps with PTFE/silicone septa, and low-volume inserts.

2.3 Instrumentation Software

1. 500 MHz (or higher) NMR spectrometer equipped with a 5 mm triple resonance probe at 300 K (or similar)—the appropriate software to control acquisition of fecal sample ^1H NMR spectra (matching, tuning, shimming), to set pulse calibration parameters, and to process.

2. Agilent 7890A GC coupled to a 5975C inert XL EI/CI MSD with triple-axis detector MS and a CTC-CH 4222 autosampler with an Agilent HP-5ms (29 m × 250 μm × 0.25 μm) in split/ splitless mode. MSD ChemStation (*Agilent Technologies, California, USA*) to acquire and process GC-MS data.

3. ACQUITY UPLC coupled to a Xevo TQD MS system (*Waters, Massachusetts, USA*) with an ACQUITY HILIC, BEH amide column (2.1 × 150 mm, 1.7 μm). Waters MassLynx® software to collect and process LC-MS/MS data.

3 Methods

3.1 Sample Collection and Storage

1. Scientists should use personal protective equipment when handling experimental animals.

2. Each animal should be placed in a separate cage during sample collection, in case where the animals are kept housed in non-individual cage (*see* **Note 6**).

3. Fecal pellets should be collected immediately after defecation to limit sample deterioration.

4. Fecal pellets should be collected using tweezers, which should be cleaned from sample to sample.

5. More than one fecal pellet should be collected, in order to subsequently receive the necessary volume of fecal extract, to pass through the filter, in sample preparation.

6. Fecal pellets should be placed in 2 mL Eppendorf tube.

7. Sodium azide could be added in order to protect fecal samples from microbial contamination.

8. Immediately after collection, the fecal samples should be frozen at $-80\ ^\circ C$, for long-term storage.

9. Fecal pellets could be subjected to snap freezing with liquid nitrogen, as this process efficiently inhibits biochemical interactions.

10. In cases where sample collection is directly from the intestine, rats should be anesthetized and sacrificed in accordance to the Helsinki Declaration for keeping and handling experimental animals.

3.2 Sample Preparation

3.2.1 Analysis Using NMR Spectroscopy

1. 500 mg of smashed stool material is placed in a 2 mL Eppendorf tube.

2. 1 mL phosphate-buffered saline (PBS 1.9 mM Na_2HPO_4, 8.1 mM NaH_2PO_4, 150 mM NaCl, pH 7.4) is added, in a ratio of 1:2 fecal sample weight to extraction buffer volume (*see* **Note 7**) [8]. PBS could be prepared either in D_2O or in distilled water. D_2O should be added at least in a ratio of 1:5 (v/v) and then made up to final volume with distilled water. NaCl is added to facilitate the extraction of fecal samples. When preparing PBS the dissolution of the salts requires vortexing and sonication.

3. Vortex-mixing is performed for 2 min.

4. The mixture is homogenized using an ultrasonic homogenizer for 15 min.

5. The fecal slurry is centrifuged for 20 min at 4 $^\circ C$ ($18,000 \times g$).

6. 400 μL of the clear supernatant is diluted with 150 μL of D_2O.

7. 50 μL of 0.1% 3-(trimethylsilyl)propionic-2,2,3,3-d4 acid sodium salt (TSP) in D_2O is added to the extract.

8. Centrifuge for 18 min at 4 °C (15,000 × g).

9. Finally 550 μL of the clear extract is placed in 5 mm NMR tube for analysis using methods such as those described in Chap. 8 (Benaki and Mikros), in the present book.

3.2.2 Analysis Using GC-MS

1. 500 mg of smashed stool material is placed in a 2 mL Eppendorf tube.

2. 1 mL of MeOH-CHCl$_3$ 1:1 (v/v) is added, in a ratio of 1:2 fecal sample weight to extraction buffer volume.

3. Vortex-mixing is applied for 2 min.

4. The mixture is homogenized by sonication for 10 min.

5. The organic fecal extract is centrifuged for 20 min at 4 °C (18,000 × g).

6. 100 μL of the supernatant is placed in a glass GC-MS vial and evaporated under a stream of nitrogen.

7. 20 μL methoxyamine hydrochloride in pyridine (40 mg/mL) and 180 μL MSTFA [1% trimethylchlorosilane (TMCS)] are added followed by incubation for 90 min at 28 °C and for 30 min at 37 °C, respectively (*see* **Note 8**).

8. 5 μL of a 1 mM solution of 2-fluorobiphenyl is added as an internal standard. An alternative option is the addition of 5 μL pentadecane (in pyridine). Before the addition of internal standard the vial should be allowed to fall to ambient temperature (*see* **Note 9**).

9. The gas chromatographic analysis (GC-MS) of the fecal extracts is performed as described in detail, in a previously published paper [8].

3.2.3 Sample Preparation for HILIC-MS

1. 250 mg of smashed stool material is placed in a 2 mL Eppendorf tube.

2. Aqueous extraction solvent, water with 1-propanol or acetonitrile (1:1 v/v), is added, in a ratio of 1:4 fecal sample weight to extraction solvent volume (*see* **Note 10**) [8, 13]. Acetonitrile extracts a great wide of metabolites, but 1-propanol preferably extracts amino acids related to bowel disease.

3. Vortex-mixing is applied for 2 min.

4. The mixture is homogenized by sonication for 10 min.

5. Ultra-centrifugation of the samples is performed 30 min at 4 °C (20,000 × g).

6. Supernatants are filtered through syringe filters (PTFE 0.22um).

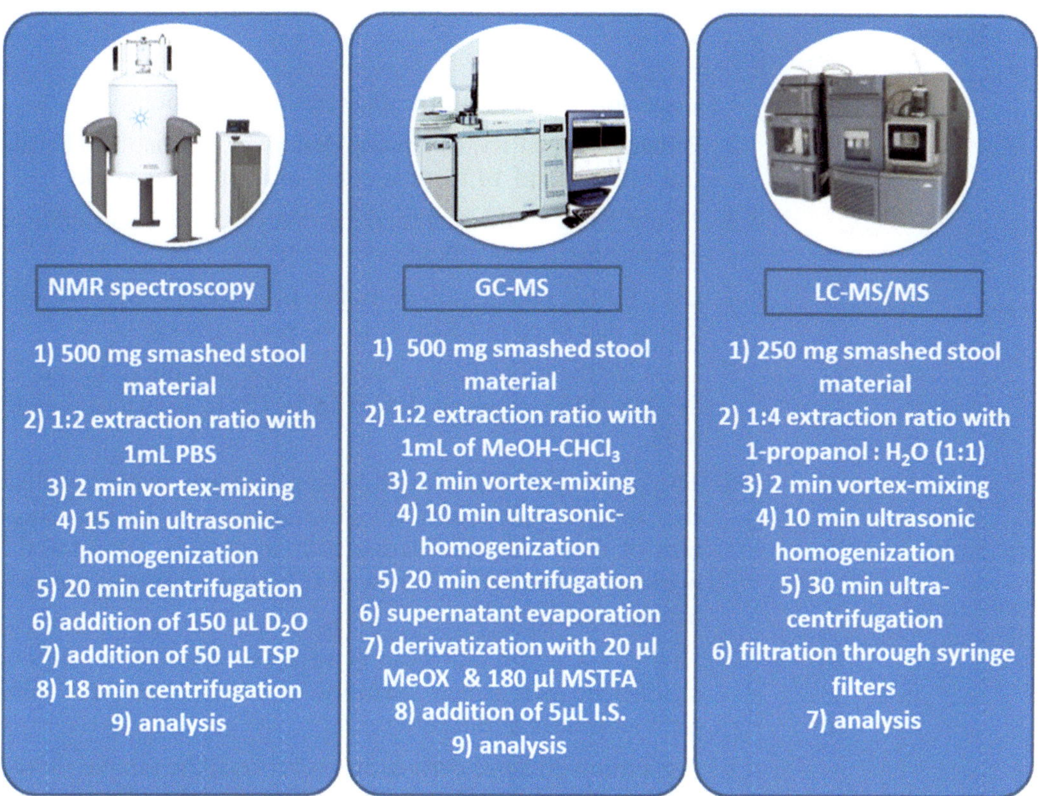

Fig. 1 Rat fecal sample preparation protocols for metabolomics-based analysis using NMR spectroscopy, GC-MS, and LC-MS/MS

7. The filtrate should be collected in an Eppendorf tube and not immediately into the LC-MS vial, which has a narrower opening.

8. The required sample volume, depending on the number of subsequent injections, is added to the LC-MS vial insert prior to analysis.

9. Analyze by targeted HILIC-MS/MS with a multi-analyte method [e.g., as the one described in Chap. 5 (Virgiliou et al.), in the present book and in a previous published paper [18]] or an untargeted method [e.g., as described in Chap. 7 (Want), in the present book].

The applied protocols for each analytical technique used are presented briefly in Fig. 1.

4 Notes

1. Used only for cleaning the analytical equipment.

2. Preparation of PBS, as already described in Chap. 8 (Benaki and Mikros) and 14 (Spyros et al), in the present book.

3. 25 mg NaCl could be added for every 10 mL PBS buffer to improve extraction efficacy.

4. Used for the preparation of mobile phases in LC-MS/MS system as described in Chap. 5, in the present book.

5. Used for the preparation of wash solvent and purge solvent of LC-MS/MS systems as described in Chap. 5, in the present book.

6. The rats in each cage should be appropriately marked so that they are effectively identified, which is necessary for the collection of their samples. Herbal hair dye can be used for the marking and should be renewed at regular intervals (approximately every 15 days).

7. Higher extraction ratios can be used (such as 1:3) in cases of insufficient fecal sample quantity, in order to ensure a sufficient aliquot of fecal extract.

8. For a more effective derivatization process, the sample should be placed in a glass insert in GC-MS glass vial.

9. Quick handling is required during the addition of the internal standard, since the volatile derivatives could be evaporated while opening the cap.

10. Lower extraction ratios can be used, such as 1:3 or 1:2, in combination with supervision of the analytical system in order to prevent problems in case of analysis of concentrated extracts.

References

1. Guinane CM, Cotter PD (2013) Role of the gut microbiota in health and chronic gastrointestinal disease: understanding a hidden metabolic organ. Ther Adv Gastroenterol 6(4):295–308

2. Holmes E, Li JV, Athanasiou T, Ashrafian H, Nicholson JK (2011) Understanding the role of gut microbiome-host metabolic signal disruption in health and disease. Trends Microbiol 19(7):349–359

3. Deda O, Gika HG, Wilson ID, Theodoridis GA (2015) An overview of fecal sample preparation for global metabolic profiling. J Pharm Biomed Anal 113:137–150

4. Theodoridis G, Gika H, Franceschi P et al (2012) LC-MS based global metabolite profiling of grapes: solvent extraction protocol optimisation. Metabolomics 8(2):175–185

5. Gika H, Theodoridis G (2011) Sample preparation prior to the LC-MS-based metabolomics/metabonomics of blood-derived samples. Bioanalysis 3(14):1647–1661

6. Gika HG, Wilson ID, Theodoridis GA (2014) LC–MS-based holistic metabolic profiling. Problems, limitations, advantages, and future perspectives. J Chromatogr B 966:1–6

7. Bollard ME, Stanley EG, Lindon JC et al (2005) NMR-based metabonomic approaches for evaluating physiological influences on biofluid composition. NMR Biomed 18(3):143–162

8. Deda O, Chatziioannou AC, Fasoula S et al (2017) Sample preparation optimization in fecal metabolic profiling. J Chromatogr B 1047:115–123

9. Hooper LV, Midtvedt T, Gordon JI (2002) How host-microbial interactions shape the nutrient environment of the mammalian intestine. Annu Rev Nutr 22:283–307

10. Lamichhane S, Yde CC, Schmedes MS et al (2015) Strategy for nuclear-magnetic-resonance-based metabolomics of human feces. Anal Chem 87(12):5930–5937

11. Wu J, An Y, Yao J, Wang Y, Tang H (2010) An optimised sample preparation method for NMR-based faecal metabonomic analysis. Analyst 135(5):1023–1030

12. Saric J, Wang Y, Li J et al (2008) Species variation in the fecal metabolome gives insight into differential gastrointestinal function. J Proteome Res 7(1):352–360

13. Deda O, Gika H, Panagoulis T et al (2017) Impact of exercise on fecal and cecal metabolome over aging: a longitudinal study in rats. Bioanalysis 9(1):21–36

14. Bezabeh T, Somorjai RL, Smith IC (2009) ICP MR metabolomics of fecal extracts: applications in the study of bowel diseases. Magn Reson Chem 47(S1):S54–S61

15. Monleon D, Garcia-Valles R, Morales JM et al (2014) Metabolomic analysis of long-term spontaneous exercise in mice suggests increased lipolysis and altered glucose metabolism when animals are at rest. J Appl Physiol 117(10):1110–1119

16. Gao X, Pujos-Guillot E, Martin J-F et al (2009) Metabolite analysis of human fecal water by gas chromatography/mass spectrometry with ethyl chloroformate derivatization. Anal Biochem 393(2):163–175

17. Gao X, Pujos-Guillot E, Sébédio J-L (2010) Development of a quantitative metabolomic approach to study clinical human fecal water metabolome based on trimethylsilylation derivatization and GC/MS analysis. Anal Chem 82(15):6447–6456

18. Virgiliou C, Sampsonidis I, Gika HG, Raikos N, Theodoridis GA (2015) Development and validation of a HILIC-MS/MS multitargeted method for metabolomics applications. Electrophoresis 36(18):2215–2225

GC-MS Metabolomic Profiling of Protic Metabolites Following Heptafluorobutyl Chloroformate Mediated Dispersive Liquid Microextraction Sample Preparation Protocol

Petr Hušek, Zdeněk Švagera, Dagmar Hanzlíková, Iva Karlínová, Lucie Řimnáčová, Helena Zahradníčková, and Petr Šimek

Abstract

A simple analytical workflow is described for gas chromatographic-mass spectrometry (GC-MS)-based metabolomic profiling of protic metabolites, particularly amino-carboxylic species in biological matrices. The sample preparation is carried out directly in aqueous samples and uses simultaneous in situ heptafluorobutyl chloroformate (HFBCF) derivatization and dispersive liquid-liquid microextraction (DLLME), followed by GC-MS analysis in single-ion monitoring (SIM) mode. The protocol involves ten simple pipetting steps and provides quantitative analysis of 132 metabolites by using two internal standards. A comment on each analytical step and explaining notes are provided with particular attention to the GC-MS analysis of 112 physiological metabolites in human urine.

Key words Metabolomic profiling, GC-MS, Dispersive liquid-liquid microextraction, Chloroformate derivatization, Urine, Quantitative analysis

1 Introduction

Comprehensive metabolomic analysis of small protic metabolites possessing amino, carboxy, thio, or hydroxy groups in complex biological matrices has been a demanding task because of the frequent occurrence of structurally close and isomeric structures that are difficult to separate and detect by at present prevailing liquid chromatographic-mass spectrometry (LC-MS) techniques. As a result, GC-MS combined with an efficient sample preparation strategy involving metabolite derivatization (as a prerequisite) has still been a popular, cost-effective tool in the analysis of amino and organic acids and other protic metabolites that play important biochemical roles in central metabolism. Current GC-MS-based metabolomics relies mainly on two derivatization strategies: (1)

Georgios A. Theodoridis et al. (eds.), *Metabolic Profiling: Methods and Protocols*, Methods in Molecular Biology, vol. 1738, https://doi.org/10.1007/978-1-4939-7643-0_11, © Springer Science+Business Media, LLC, part of Springer Nature 2018

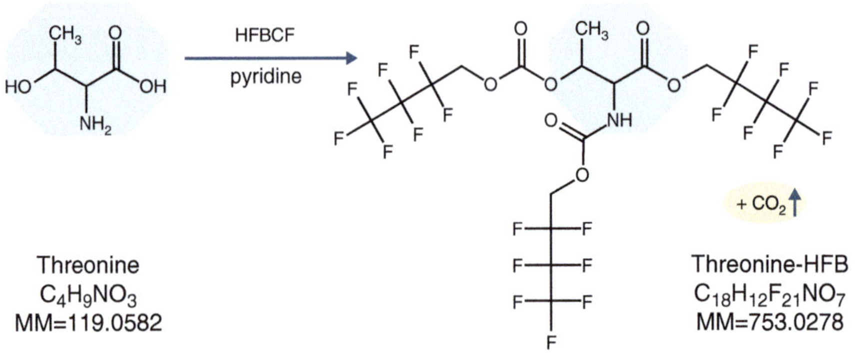

HFBCF

pyridine

Threonine
C₄H₉NO₃
MM=119.0582

+ CO₂

Threonine-HFB
C₁₈H₁₂F₂₁NO₇
MM=753.0278

Fig. 1 Reaction scheme for the threonine protic functional groups with the HFBCF reagent. The carboxyl group yields a HFB ester and the amino group a corresponding HFB carbamate, while the hydroxyl is transformed into a HFB carbonate. *MM* = Monoisotopic mass

oximation with silylation and (2) reaction with alkyl chloroformates. The former approach requires strictly anhydrous condition for silylation and has proved useful for profiling of polyhydroxylic metabolites such as sugars [1, 2], steroids [3], sterols, and tocopherols [4] but much less effective for metabolites bearing protic nitrogen functional groups [5–7]. The latter approach has been complementary and even more attractive; it can be applied in situ in a complex biological matrix, and liquid-liquid microextraction proceeds simultaneously under pyridine catalysis and, importantly, with simultaneous carbon dioxide evolvement [8–11]. The arising CO_2 is partly dissolved in the whole sample medium; it enhances the effective surface area between the immiscible organic and aqueous phase and renders thus a powerful dispersive phase enabling the system to reach the final equilibrium in 5 s [4, 12].

Fluoroalkyl chloroformates (FCFs) possess some advanced features over the traditionally used alkyl chloroformates (RCFs) providing highly volatile and much less polar derivatives extractable into nonpolar hydrocarbon solvents. The reaction scheme for reaction of heptafluorobutyl chloroformate (HFBCF) with threonine, an amino acid possessing 3 protic functional groups, is shown in Fig. 1.

The reaction product is formed directly in an aqueous environment with high yields in less than 5 s. In this way, extraordinary clean protic metabolite extracts have been obtained from complex biological matrices and have been successfully applied to metabolite profiling of amino acids and steroids in human serum [4, 13] or amino-carboxylic metabolites in human urine [12]. Moreover, the clean extracts enabled chiral GC-MS analysis of 35 amino acid enantiomeric pairs in human serum [14].

The workflow of the GC-MS-based metabolomic analysis is depicted in Fig. 2.

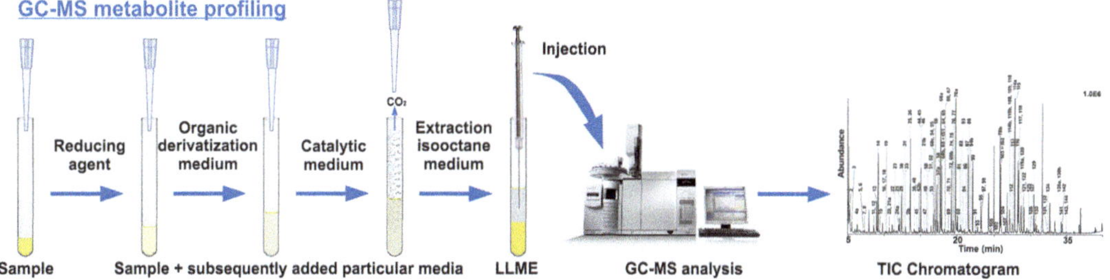

Fig. 2 The workflow for the GC-MS metabolomic analysis of protic metabolites in aqueous biological matrices

Here, we describe a HFBCF-based GC-MS profiling method for quantification of protic metabolites possessing amino, carboxy, activated hydroxy, and thiol functional groups [12]. The elaborated sample preparation protocol is simple and fast and follows a procedure described in detail for urinary analysis in reference [12] which should be consulted whenever necessary for getting more comprehensive knowledge. It involves gradual pipetting of uniform small volumes of a sample and necessary liquid media in ten steps:

(1) A sample

(2) An internal standard solution

(3) A reducing medium

(4) A pH adjustment

(5) An organic reaction medium containing the HFBCF reagent

(6 and 7) A repeated addition of a catalytic medium with pyridine

(8) An organic extraction medium

(9) An acidification medium

(10) An upper extraction phase transfer into a GC autosampler vial (*see* Fig. 3) and, finally, the sample extract injection into a GC-MS spectrometer.

Although the protocol was primarily developed for metabolomic GC-MS analysis of protic metabolites in urine normalized to creatinine [12], the procedure can directly be applied to any aqueous biological material with low protein content. If the content of biopolymers is high (>2 mg/mL), then a prior precipitation step must be adapted to the described workflow (*see* **Note 1**). The protocol may be modified for some applications, for instance, the reducing step 3 can be omitted without any change, if the reduction of disulfide bonds is not required.

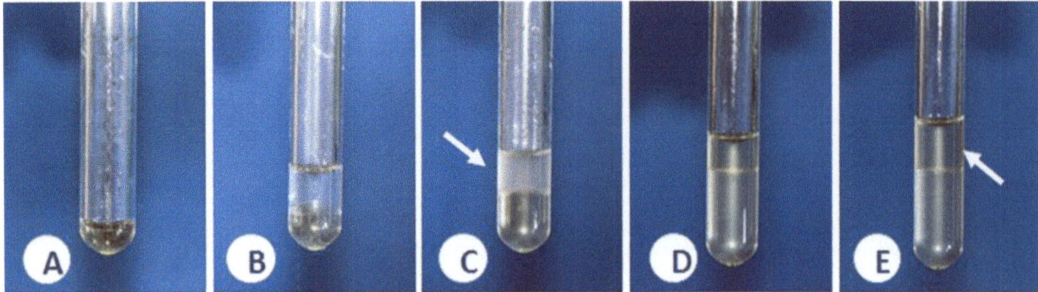

Fig. 3 A view on a 6 × 50 mm culture tube containing an aqueous sample (here urine) or an aqueous sample extract and gradually added media during the sample preparation process: (**a**) an aqueous sample; (**b**) an arising two-phase system after performing steps 2–5, before the reaction initiation; (**c**) a turbid upper phase after the first addition of the catalytic medium, step 6; occasionally visible CO2 bubbles can appear; (**d**) the organic upper phase is clarified after step 7, which indicates that the reaction was completed; (**e**) increasing the sample extract volume and its acidification in steps 8–9 enables an easy organic upper phase into an autosampler vial in step 10 and final GC-MS analysis

2   Materials

2.1   Samples

Samples containing no or little protein and cell residues (urine, cell culture media) or cell and tissue extracts.

2.2   Chemicals, Solutions, and Reaction Media

1. Chemicals of analytical grade should be used. All solutions, except those containing the HFBCF reagent, should be prepared in distilled deionized water (DI water, <1.5 μS/cm, 25 °C) and stored at 4 °C (unless otherwise indicated).

2. Tris(3-hydroxypropyl)phosphine reducing agent (THP, 80%, Merck): prepare a 5% stock solution by transferring 0.625 mL of the THP liquid to a 10 mL volumetric flask, and adjust by DI water up to 10 mL. Prepare a 0.5% working solution by dilution 1:9 in a 10 mL volumetric flask (*see* **Note 2**).

3. 100 mM $NaHCO_3$ (99.998% purity, Alfa Aesar) solution: dissolve 840 mg in 100 mL of DI water.

4. 1 M NaOH (99.99% purity, Alfa Aesar) solution: dissolve 4 g in 100 mL of DI water.

5. Heptafluorobutanol (HFBOH) (99% purity, Pragolab, Prague, Czech Republic).

6. Heptafluorobutyl chloroformate (HFBCF, 98%, Pragolab) (*see* **Note 3**).

7. Solvents: isooctane (2,2,4-trimethylpentane, 99.5%), pyridine (p.a., 99.0%), isopropanol (2-propanol, 99.5% purity) (all Sigma-Aldrich).

8. The organic reaction medium: prepare isooctane, HFBCF, and HFBOH in a volume ratio 15:4:1 (v/v/v) in a

Teflon-capped, well-tightened 4 mL glass vial. Store in a refrigerator, where the mixture remains stable for several months.

9. The catalytic medium: mix 1 M NaOH with pyridine in a volume ratio of 24:1 (v/v).

10. The artificial urine solution: prepare the following chemicals in DI water to final 10 g/L urea, 1 g/L creatinine, 7 g/L NaCl, and 3 g/L K_2SO_4 [12].

11. The certified urine standard (product ORG-01) containing diagnostic organic acids (ERNDIM Foundation (http://cms.erndimqa.nl/). For the analyte concentrations, refer to the website. Order the latest available batch.

2.3 Analytical Standards and Stock Solutions

1. Internal standard solution (IS): 4-phenylbutyric acid (4PB, Sigma-Aldrich, MW = 164.2 g/mol); homophenylalanine (hF, Sigma-Aldrich, MW = 179.2 g/mol). Prepare a stock solution in 100 mM $NaHCO_3$ with a final concentration of 200 µmol/L, i.e., 5 nmol in 25 µL of the applied internal standard solution.

2. Protein amino acid (AA) standard solution (Sigma-Aldrich, P/N AAS-18) in 0.1 M HCl containing alanine, glycine, valine, leucine, isoleucine, threonine, serine, proline, aspartic acid, methionine, glutamic acid, phenylalanine, lysine, histidine, tyrosine, and cystine at a concentration of 2.5 and 1.25 mmol/L, respectively. Alternatively, the protein AA mixture can be prepared from stock solutions of particular AAs in 0.1 M HC. For the complete metabolite list, refer to Table 1.

3. Non-protein amino acid, biogenic amine (Sigma-Aldrich) standard solutions: prepare a stock solution of each metabolite in 100 mM HCl with a final concentration of 100 µmol/L (Table 1). Store at 4 °C.

4. Organic acid standards (Sigma-Aldrich); less common carboxylic acids 2-hydroxysebacic, 3-hydroxyadipic, 3-hydroxypropionic, 3-hydroxyvaleric, and 5-hydroxyhexanoic acid and glycine conjugates hexanoylglycine, methylcrotonylglycine, and tiglylglycine can be purchased from Dr. E. Brunet, Dept. Organic Chemistry, University Autonoma de Madrid (Madrid, Spain), http://www.uam.es/gruposinv/lumila/list.pdf; a racemate of 2-methyl citric acid (90%, C/D/N Isotopes, P/N X-4176).

5. The organic acid stock solutions data are summarized in Table 1 (*see* **Note 4**).

2.4 GC-MS Instrumentation

1. Agilent 7890A GC system equipped with G4513A autosampler (Agilent), multimode injector (MMI), equipped with a 10 µL syringe (CTC Analytics, P/N PAL3-SYH-207807).

Table 1
The list of protic metabolites determined by the described GC-MS protocol in human urine. The traditional name, the metabolite product after the HFBCF derivatization, retention data, monoisotopic mass of the arising HFBCF derivatized product, diagnostic SIM ions, the used internal standard, the observed medium metabolite concentration in urine [12]

No.	No. [12][a]	Traditional name	The profiled metabolites as the HFBCF derivatives	GC RT (min)	MM product	m/z^{q1}	m/z^{q2}	Internal Standard[b]	L3 c (μM)[c]	Stock solution	Stock solution c (mM)	HMDB	PubChem	KEGG
1	108	1-Methylhistidine	1-Methylhistidine	27.2	577.1	95	350	hF	40	0.1 M HCl	10	HMDB00001	92105	C01152
2	101	2,4-Diamino butyric acid	2,4-Diamino butyrate	24.74	752.0	282	256	hF	40	DI water	10	HMDB02362	470	
3	36	2-Aminobutyric acid	2-Amino butyrate	13.61	511.0	284	84	hF	20	DI water	10	HMDB00452	80283	C02356
4	99	2-Amino heptanedioic-acid	2-Aminopimelate	24.6	765.1	338	138	hF	40	DI water	10	HMDB34252	101122	
5	27	2-Aminoiso-butyric acid	2-Aminoiso butyrate	12.14	525.1	284	241	hF	40	DI water	50	HMDB01906	6119	C03665
6	22	2-Hydroxy-3-methylbutyric acid	2-Hydroxy-3-methylbutyrate (isovalerate)	11.16	526.0	55	299	PB	20	DI water	10	HMDB00407	99823	
7	34	2-Hydroxy-3-methylpentanoic acid	2-Hydroxy-3-methylvalerate	13.06	540.1	284	484	PB	20	DI water	10	HMDB00317	10796774	

8	19	2-Hydroxybutyric acid	2-Hydroxybutyrate	10.49	512.0	285	241	PB	20	DI water	10	HMDB00008	11266	C05984
9	112	2-Hydroxy decanedioic acid	2-Hydroxysebacate	27.34	808.1	95	381	PB	20	Acetonitrile	10	HMDB00424	128458	
10	65	2-Hydroxyglutaric acid	2-Hydroxyglutarate	18.34	738.0	283	239	PB	100	DI water	100	HMDB02307	439340	C00894
11	11	2-Hydroxyisobutyric acid	2-Hydroxyisobutyrate	8.74	512.0	241	285	PB	40	DI water	10	HMDB00729	11671	
12	82	2-Hydroxyphenylacetic acid	2-Hydroxyphenylacetate	20.52	560.0	91	333	PB	20	DI water	50	HMDB00669	11970	C05852
13	26	2-Hydroxyvaleric acid	2-Hydroxyvalerate	12.12	526.0	55	299	PB	20	DI water	10	HMDB01863	98009	
14	3	2-Ketoisovaleric acid	2-Ketoisovalerate	5.81	298.0	71	113	PB	20	DI water	10	HMDB00019	49	C00141
15	68a	2-Methylcitric acid	2-Methylcitrate-4 (lactone)	18.46	552.0	152	334	PB	40	DI water	10	HMDB00379	515	
16	46	3-Aminoisobutanoic acid	3-Aminoisobutyrate	15.02	511.0	256	112	hF	100	0.1 M HCl	100	HMDB03911	64956	C05145
17	67	3-Hydroxyadipic acid	3-Hydroxyadipate	18.41	752.0	85	127	PB	100	0.1 M HCl	50	HMDB00345	151913	
18	24a	3-Hydroxybutyric acid	3-Hydroxybutyrate-2	11.93	512.0	268	69	PB	100	DI water	10	HMDB00357	441	C01089
19	4a	3-Hydroxyisovaleric acid	3-Hydroxyisovalerate-1(OH)	6.46	300.1	59	85	PB	100	DI water	10	HMDB00754	69362	

(continued)

Table 1
(continued)

No.	No. [12][a]	Traditional name	The profiled metabolites as the HFBCF derivatives	GC RT (min)	MM product	Diagnostic ions m/z^{q1}	m/z^{q2}	Internal Standard[b]	L3 c (μM)[c]	Stock solution	Stock solution c (mM)	Metabolite database coding HMDB	PubChem	KEGG
20	41	3-Hydroxy methylglutaric acid	3-Hydroxy-3-methyl glutarate	14.37	752.0	85	285	PB	100	DI water	10	HMDB00355	1662	C03761
21	91	3-Hydroxy phenylacetic acid	3-Hydroxyphenyl acetate	22.61	560.0	333	277	PB	20	0.1 M HCl	100	HMDB00440	12122	C05593
22	89a	3-Hydroxyproline	3-Hydroxyproline-2	21.9	765.0	521	538	hF	20	0.1 M HCl	10	HMDB02113	11137200	C04397
23	79a	3-Hydroxysebacic acid	3-Hydroxysebacate-1(OH)	20.01	582.1	71	271	PB	20	DI water	10	HMDB00350	3017884	
24	9a	3-Hydroxyvaleric acid	3-Hydroxyvalerate-1(OH)	8.21	300.1	71	271	PB	40	DI water	10	HMDB00531	107802	
25	131	3-Methoxytyramine	3-Methoxytyramine	32.36	845.1	319	376	hF	20	0.1 M HCl	10	HMDB00022	1669	C05587
26	5	3-Methyl-2-Oxovaleric acid	2-Keto-3-methylvalerate	7.34	312.1	57	85	PB	20	DI water	10	HMDB00491	47	C03465
27	47	3-Methyladipic acid	3-Methyladipate	15.53	524.1	325	55	PB	20	DI water	50	HMDB00555	6999745	
28	76	3-Methylcrotonyl glycine	Methylcrotonyl glycine	19.54	339.1	83	82	hF	40	DI water	10	HMDB00459	169485	
29	116	3-Methylhistidine	3-Methylhistidine	28.38	577.1	95	150	hF	40	0.1 M HCl	10	HMDB00479	64969	C01152
30	111	4-Aminobenzoic acid	4-Aminobenzoate	27.29	545.0	146	345	hF	20	Ethanol	10	HMDB01392	978	C00568

31	87	4-Hydroxybenzoic acid	4-Hydroxybenzoate	21.41	546.0	303	347	PB	20	Ethanol	100	HMDB00500	135	C00156
32	38	4-Hydroxybutyric acid	4-Hydroxybutyrate	14.05	512.0	227	269	PB	100	DI water	100	HMDB00710	10413	C00989
33	115	4-Hydroxycinnamic acid	4-Hydroxycinnamate	27.98	572.0	329	572	PB	20	DI water	10	HMDB02035	637542	C00811
34	110	4-Hydroxymandelic acid	4-Hydroxymandelate	27.25	802.0	575	347	PB	20	DI water	10	HMDB00822	328	C11527
35	96	4-Hydroxy phenylacetic acid	4-Hydroxy phenylacetate	23.31	560.0	289	333	PB	40	0.1 M HCl	100	HMDB00020	127	C00642
36	94a	4-Hydroxyproline	4-Hydroxyproline-2	23.12	765.0	294	521	hF	20	0.1 M HCl	50	HMDB06055	69248	C01015
37	63	4-Phenylbutyric acid (4 PB, I.S.)	4-Phenylbutyrate	17.95	346.1	104	147			0.1 M NaHCO3	100			
38	97	5-Aminolevulinic acid	5-Aminolevulinate	23.36	539.0	283	256	hF	40	0.1 M HCl	10	HMDB01149	137	C00430
39	81	5-Aminopentanoic acid	5-Aminovalerate	20.41	525.1	256	269	hF	40	0.1 M HCl	10	HMDB03355	138	C00431
40	53	5-Hydroxyhexanoic acid	5-Hydroxyhexanoate	16.67	540.1	227	113	PB	100	DI water	10	HMDB00525	170748	
41	144	5-Hydroxyindo leacetic acid	5-Hydroxyin doleacetate	35.16	599.0	372	599	hF	20	DI water	10	HMDB00763	1826	C05635
42	138b	5-Hydroxylysine	5-Hydroxylysine (isomers)	34.12	1022.1	269	256	hF	40	0.1 M HCl	50	HMDB00450	3032849	C16741

(continued)

Table 1
(continued)

| No. | No. [12][a] | Traditional name | The profiled metabolites as the HFBCF derivatives | GC | | Diagnostic ions | | Internal | L3 | Stock | Stock solution | Metabolite database coding | | |
				RT (min)	MM product	m/z^{q1}	m/z^{q2}	Standard[b]	c $(\mu M)^c$	Stock solution	c (mM)	HMDB	PubChem	KEGG
43	37a	Acetylglycine	N-Acetylglycine-2(NH)	13.88	525.0	256	483	hF	200	DI water	10	HMDB00532	10972	
44	54	Aconitic acid *(trans)*	Aconitate	16.78	720.0	321	492	PB	200	DI water	100	HMDB00958	444212	C02341
45	45	Adipic acid	Adipate	14.95	510.1	282	311	PB	40	Ethanol	100	HMDB00448	196	C06104
46	29	Alanine	Alanine	12.35	497.0	270	70	hF	40	0.1 M HCl	100	HMDB00161	5950	C00041
47	93	Aminoadipic acid	2-Aminoadipate	22.88	751.0	124	282	hF	40	0.1 M HCl	100	HMDB00510	469	C00956
48	70	Asparagine	Asparagine	19.1	522.0	295	95	hF	100	0.1 M HCl	100	HMDB00168	6267	C00152
49	64	Aspartic acid	Aspartate	18.3	723.0	254	496	hF	40	0.1 M HCl	100	HMDB00191	5960	C00049
50	86	Azelaic acid	Azelaate	21.23	552.1	353	152	PB	20	Ethanol	10	HMDB00784	2266	C08261
51	20	Benzoic acid	Benzoate	10.74	304.0	105	304	PB	40	Ethanol	100	HMDB01870	243	C00180
52	44	Beta-alanine	3-alanine	14.91	497.0	270	113	hF	40	0.1 M HCl	10	HMDB00056	239	C00099
53	16	Citraconic acid	Citraconate	10.29	494.0	295	267	PB	40	DI water	10	HMDB00634	643798	C02226
54	21a	Citramalic acid	Citramalate-1 (lactone)	10.8	312.0	85	285	PB	100	DI water	10	HMDB00426	1081	C00815
55	60a	Citric acid	Citrate-2 (OH)	17.79	738.0	311	269	PB	400	DI water	100	HMDB00094	311	C00158

56	132	Cystathionine	Cystathionine	32.42	1038.0	328	282	hF	40	0.1 M HCl	100	HMDB00099	439258	C02291
57	90	Cysteine	Cysteine (total)**	22.22	755.0	328	285	hF	100	0.1 M HCl	100	HMDB00574	5862	C00097
58	124a	Diaminopimelic acid	2,6-Diaminopimelate	29.45	1006.1	308	536	hF	40	DI water	10	HMDB01370	439283	C00666
59	135	DOPA	3,4-Dihydroxy phenyl alanine	32.63	1057.0	149	388	hF	40	0.1 M HCl	10	HMDB00609	836	C00355
60	12	Ethylmalonic acid	Ethylmalonate	8.78	496.0	297	468	PB	40	DI water	10	HMDB00622	11756	
61	130	Ferulic acid (*trans*)	4-Hydroxy-3-methoxycinnamate	30.64	602.0	602	375	PB	20	Ethanol	100	HMDB00954	445858	C01494
62	15	Fumaric acid	Fumarate	9.53	480.0	281	253	PB	40	DI water	10	HMDB00134	444972	C00122
63	83	Glutamic acid	Glutamate	20.82	737.0	310	282	hF	40	0.1 M HCl	10	HMDB03339	23327	C00217
64	107	Glutamine	Glutamine	26.64	554.1	84	282	hF	100	DI water	100	HMDB00641	5961	C00064
65	30	Glutaric acid	Glutarate	12.51	496.0	227	297	PB	40	DI water	10	HMDB00661	743	C00489
66	58	Glyceric acid	Glycerate (2,3-Dihydroxy propionate)	17.68	740.0	113	497	PB	40	DI water	10	HMDB00139	439194	C00258
67	35	Glycine	Glycine	13.56	483.0	256	212	hF	200	DI water	100	HMDB00123	750	C00037
68	14	Glycolic acid	Glycolate	9.35	484.0	285	213	PB	40	DI water	100	HMDB00115	757	C00160
69	125	Glycylproline	Glycylproline	29.83	580.1	70	153	hF	40	0.1 M HCl	10	HMDB00721	79101	
70	39a	Hexanoylglycine	Hexanoylglycine-1 (cyclic)	14.07	155.1	99	71	hF	20	DI water	10	HMDB00701	99463	

(continued)

Table 1
(continued)

No.	No. [12][a]	Traditional name	The profiled metabolites as the HFBCF derivatives	GC RT (min)	MM product	Diagnostic ions m/z^{q1}	m/z^{q2}	Internal Standard[b]	L3 c (μM)[c]	Stock solution	Stock solution c (mM)	Metabolite database coding HMDB	PubChem	KEGG
71	78a	Hippuric acid	Hippurate-1 (cyclic, 60%)	19.86	161.0	105	161	hF	400	Ethanol	100	HMDB00714	464	C01586
72	106	Histamine	Histamine	26.48	563.1	308	320	hF	40	0.1 M HCl	10	HMDB00870	774	C00388
73	114a	Histidine	Histidine-2 (NR)	27.9	789.0	307	362	hF	200	0.1 M HCl	100	HMDB00177	6274	C00135
74	100	Homocysteine	Homocysteine (total)**	24.68	769.0	282	342	hF	40	MeCN	10	HMDB00742	778	C05330
75	103	Homophenylalanine (hF, I.S.)	Homophenylalanine	25.98	587.1	91	283			0.1 M NaHCO$_3$	100			
76	104	Homovanillic acid	Homovanillate	26.07	590.0	107	590	PB	20	0.1 M HCl	50	HMDB00118	1738	C05582
77	32	Hydroxyisocaproicacid	2-Hydroxyisocaproate	12.79	540.1	296	113	PB	20	DI water	10	HMDB00746	83697	
78	118	Hydroxyphenyllactic acid	4-hydroxy-Phenyllactate	28.54	816.0	572	345	PB	20	DI water	10	HMDB00755	9378	C03672
79	52	Hydroxypropionic acid	3-Hydroxypropionate (dimer)	16.06	526.0	255	298	PB	100	DI water	10	HMDB00700	68152	C01013
80	113	Indolacetate	Indolacetate	27.61	357.1	130	357	hF	20	MeCN	10			
81	80	Isocitric acid	Isocitrate	20.18	964.0	465	321	PB	100	DI water	100	HMDB00193	1198	C00311
82	50	Isoleucine	Isoleucine	15.93	539.1	283	312	hF	40	0.1 M HCl	10	HMDB00172	6306	C00407

83	59	Isovalerylglycine	Isovalerylglycine	17.77	341.1	85	525	hF	40	DI water	10	HMDB00678	546304	
84	6	Ketoleucine	2-Ketoisocaproate	7.43	312.1	85	57	PB	20	DI water	10	HMDB00695	70	C00233
85	126	Kynurenic acid	Kynurenate	29.99	597.0	371	354	hF	40	NaHCO3	10	HMDB00715	3845	C01717
86	143	Kynurenine	Kynurenine	35.07	842.1	146	372	hF	40	0.1 M HCl	10	HMDB00684	161166	C00328
87	13	Lactic acid	Lactate	9.14	498.0	271	255	PB	100	DI water	100	HMDB00190	107689	C00186
88	49	Leucine	Leucine	15.84	539.1	312	270	hF	40	0.1 M HCl	50	HMDB00687	6106	C00123
89	119a	Lysine	Lysine-2 (*N*,*N*-R)	28.79	780.1	310	256	hF	100	0.1 M HCl	100	HMDB00182	5962	C00047
90	51	Malic acid	Malate	16.04	724.0	281	253	PB	40	DI water	10	HMDB31518	92824	C00497
91	7	Malonic acid	Malonate	7.51	468.0	269	407	PB	40	Ethanol	10	HMDB00691	867	C00383
92	74	Mandelic acid	Mandelate	19.45	560.0	289	333	PB	20	DI water	10	HMDB00703	439616	C01984
93	88	Methionine	Methionine	21.65	557.0	61	357	hF	40	0.1 M HCl	10	HMDB00696	6137	C00073
94	117	Methioninesulfone	Methionine sulfone	28.51	589.0	282	82	hF	40	0.1 M HCl	10			
95	71	Methylcysteine	*S*-Methylcysteine	19.13	543.0	61	300	hF	20	0.1 M HCl	10	HMDB02108	24417	
96	33	Methylglutaric acid	3-Methylglutarate	12.93	510.1	311	282	PB	40	DI water	10	HMDB00752	12284	
97	8	Methylmalonic acid	Methylmalonate	7.54	482.0	283	438	PB	40	DI water	10	HMDB00202	487	C02170
98	18	Methylsuccinic acid	Methylsuccinate	10.38	496.0	297	268	PB	40	DI water	50	HMDB01844	10349	C08645
99	66	*N*-Acetyl-asparticacid	*N*-Acetylaspartate	18.38	539.0	270	312	hF	200	0.1 M HCl	50	HMDB00812	65065	C01042
100	28	Nicotinic acid	Nicotinate	12.17	305.0	106	78	hF	40	0.1 M HCl	10	HMDB01488	938	C00253

(continued)

Table 1
(continued)

No.	No. [12][a]	Traditional name	The profiled metabolites as the HFBCF derivatives	GC RT (min)	MM product	Diagnostic ions m/z^{q1}	m/z^{q2}	Internal Standard[b]	L3 c (μM)[c]	Stock solution	Stock solution c (mM)	Metabolite database coding HMDB	PubChem	KEGG
101	109	Ornithine	Ornithine	27.25	766.1	296	256	hF	40	0.1 M HCl	50	HMDB00214	6262	C00077
102	2	Oxalic acid	Oxalate	5.38	454.0	113	183	PB	200	DI water	10	HMDB02329	971	C00209
103	42b	Oxoglutaric acid	2-Ketoglutarate-2(80%)	14.6	510.0	283	284	PB	200	DI water	100	HMDB00208	51	C00026
104	121	Palmitic acid	Palmitate	28.91	438.2	255	438	PB	20	Ethanol	50	HMDB00220	985	C00249
105	31	Phenylacetic acid	Phenylacetate	12.65	318.0	91	318	PB	20	Ethanol	10	HMDB00209	999	C07086
106	98	Phenylalanine	Phenylalanine	23.39	573.1	91	330	hF	40	0.1 M HCl	100	HMDB00159	6140	C00079
107	84	Phenyllactic acid	3-Phenyllactate	20.92	574.0	330	131	PB	20	DI water	10	HMDB00779	3848	C01479
108	92	Phenylpyruvic acid	Phenylpyruvate	22.84	572.0	118	329	PB	40	Ethanol	10	HMDB00205	997	C00166
109	61	Phthalic acid	Phthalate	17.79	530.0	331	332	PB	20	DI water	10	HMDB02107	1017	C01606
110	56	Pimelic acid	Pimelate	17.09	524.1	296	325	PB	20	Ethanol	10	HMDB00857	385	C02656
111	55	Proline	Proline	16.84	523.0	296	297	hF	40	0.1 M HCl	10	HMDB00162	145742	C00148
112	142	Prolylhydroxyproline	Prolylhydroxy proline	34.6	862.1	296	297	hF	100	0.1 M HCl	10	HMDB06695	11902892	
113	48	Propionylglycine	Propionylglycine	15.55	313.1	57	56	hF	100	DI water	10	HMDB00783	98681	
114	10	Propyl pentanoate	2-Propylvalerate	8.61	326.1	255	284	PB	40	DI water	10	HMDB40296	67328	
115	69	Pyroglutamic acid	Pyroglutamate	18.9	537.0	310	84	hF	200	DI water	100	HMDB00267	7405	C01879
116	73	Salicylic acid	Salicylate	19.41	546.0	120	303	PB	20	DI water	10	HMDB01895	338	C00805

117	120	Salicyluric acid	2-Hydroxy hippurate	28.82	603.0	120	403	hF	40	0.1 M HCl	10	HMDB00840	10253	C07588
118	25	Sarcosine	Sarcosine	12.07	497.0	270	226	hF	20	0.1 M HCl	10	HMDB00271	1088	C00213
119	95	Sebacic acid	Sebacate	23.16	566.1	98	367	PB	40	Ethanol	10	HMDB00792	5192	C08277
120	77	Serine	Serine	19.6	739.0	268	295	hF	100	DI water	100	HMDB00187	5951	C00065
121	134	Stearic acid	Stearate	32.5	466.3	255	466	PB	20	Ethanol	50	HMDB00827	5281	C01530
122	72	Suberic acid	Suberate	19.2	538.1	339	138	PB	20	MeCN	10	HMDB00893	10457	C08278
123	17	Succinic acid	Succinate	10.32	482.0	283	55	PB	40	DI water	100	HMDB00254	1110	C00042
124	75	Thioproline	Thioproline	19.46	541.0	314	287	hF	20	0.1 M HCl	10			
125	57a	Threonine	Threonine-1(OH)	17.42	527.0	100	283	hF	100	0.1 M HCl	100	HMDB00167	6288	C00188
126	23	Tiglylglycine	Tiglylglycine	11.27	339.1	83	55	hF	40	DI water	50	HMDB00959	6441567	
127	139	Tryptamine	Tryptamine	34.18	612.1	130	386	hF	20	0.1 M HCl	10	HMDB00303	1150	C00398
128	141	Tryptophan	Tryptophan	34.44	612.1	130	131	hF	40	NaOH	100	HMDB00929	6305	C00078
129	127	Tyramine	Tyramine	30.06	589.1	346	333	hF	20	Ethanol	10	HMDB00306	5610	C00483
130	129	Tyrosine	Tyrosine	30.58	815.0	333	289	hF	40	0.1 M HCl	100	HMDB00158	6057	C00082
131	102	Urocanic acid	Trans-urocanate	25.39	546.0	347	546	hF	20	DI water	10	HMDB34174	1549103	
132	40	Valine	Valine	14.1	525.1	298	283	hF	40	0.1 M HCl	100	HMDB00883	6287	C00183
133	128	Vanillactic acid	4-Hydroxy-3-methoxy phenyllactate	30.38	846.0	375	561	PB	20	DI water	10	HMDB00913	160637	
134	122	Vanillylmandelic acid	Vanillylmandelate	28.92	832.0	832	377	PB	20	0.1 M HCl	10	HMDB00291	736172	C05584

MM = a molecular mass of each observed metabolite derivative; *m/z* = diagnostic ions in the EI spectrum of each metabolite; q1, q2 = diagnostic (quantitation and qualifier) fragment ions used for the quantitative GC-SIM-EI-MS analysis

[a]Numbering, traditional name, and metabolite product names according to the reference 12

[b]Calibration against the internal standard; *hF* homophenylalanine, *PB* 4-phenylbutyric acid

[c]Calibration level L3 (the observed average concentration in urine) [12]

2. Sky® 4 mm I.D. cyclo double taper inlet liner (Restek, P/N 23310).

3. ZB-XLB type, 30 m × 0.25 mm ID, 0.25 μm film thickness (Phenomenex, P/N 7HG-G019-11).

4. Single quadrupole mass triple-axis detector (5975 MSD Inert XL, Agilent) equipped with an inert EI ion source.

5. Autosampler 2 mL vials (12 × 32 mm) with 9 mm PP open hole caps (Labicom, P/N 5310F-09) and 0.040″ PTFE/silicone/PTFE Septa (Labicom, P/N 604060-09).

6. Inert conical glass insert, 200 μL volume (Chromacol, P/N 02-MTV).

2.5 Additional Equipment

1. Sample preparation glass culture tubes 6 × 50 mm; material: sodium-potassium silicate

2. Glass (Merci, P/N Z1632000605010) or borosilicate glass (Kimble-Kontes, P/N 73500-650).

3. Common screw cap Teflon-lined 2 and 4 mL amber vials for the reagent solutions.

4. An adjustable 50 and 100 μL Transferpettor pipette with a glass capillary (Brand, P/N 701868 and 701873) for manipulation with the reagents and their mixtures in isooctane. The pipette tips with 25 mm capillary (gel-loading type, VWR Int.) for aspirating the upper organic phase in the 6 × 50 mm vial.

5. Alternatively, a common pipette (10–100 μL) such as a Biohit Proline® mechanical pipette (Sartorius, P/N 720050) equipped with an Optifit tip 200 (P/N 4059.9002) can be used.

6. A common vortex for sample mixing and a minicentrifuge such as mySPIN 6 (Thermo Scientific, 2000 × g) for a complementary separation of immiscible layers in sample vials.

7. A commercial assay kit for creatinine analysis. For instance, a creatinine kit (Dialab, P/N D95595).

8. A common spectrophotometer capable of measuring creatinine at 490–510 nm in urine, such as Specord® Plus (Jena Analytik), by an appropriate kit (*see* Subheading 2.5, **item** 7) used for clinical applications.

3 Methods

3.1 Sampling and Storage and Normalization of Samples

Using appropriate laboratory wear, glasses, and other personal protective equipment is recommended; follow standard laboratory precautions and local guidelines, especially waste disposal regulations. Use a functional fume hood for the sample workup.

1. Serious attention should be paid to sample collection, transport, and storage, because any omission may result in false results. For urine, collection of the morning second-void samples is the most common practice and the easiest sampling method.

2. For urine analysis, store freshly collected samples at 4 °C within 2 h. For longer than 48 h storage, freeze the samples and keep at −20 °C (*see* **Note 5**).

3. If urine is a subject of study, measure creatinine concentration in each collected urine aliquot used in the analysis data for normalization of each determined metabolite (*see* **Note 6**).

3.2 Preparation of Calibration Solutions

1. Prepare a calibration mixture following the procedure described in ref. 12 by adding an appropriate volume of each stock solution standard (Table 1) into a volumetric flask. Adjust to a final volume with the artificial urine solution.

2. The concentration of each individual analyte denotes its average level (here, level 3 = L3) measured in a pooled urine sample of normal morning urine, Table 1 [12]. Prepare the lower (L1 and L2) and higher (L4, L5) calibration points, i.e., 10 times, 2.5 times diluted, and 2.5 times and 10 times increased to the respective medium L3 level. For the metabolites with highest abundance, prepare level L6 (25 times higher than L3).

3. Distribute the calibration solution into appropriate aliquots before freezing.

4. Prepare an appropriate pooled sample by mixing an equal small volume of all samples included in the study for the verification of the average EI MS response for each target metabolite and for the quality control (QC) analysis (see also Chapter 2).

3.3 Sample Preparation Protocol

1. Transfer 25 μL of aqueous sample into a 6 × 50 mm culture tube.

2. Spike the sample with 25 μL of the internal standard solution.

3. Add 25 μL 0.5% THP reducing solution, and mix the contents gently for 1–2 s and leave to stand for 1 min.

4. Adjust pH to ca 9 with 25 μL 100 mM $NaHCO_3$ solution and vortex gently.

5. Add 50 μL of the organic reaction medium (isooctane, HFBCF, and HFBOH, 15:4:1, v/v/v) (*see* **Note 7**).

6. Add 25 μL of the catalytic medium (1 M NaOH-pyridine, 24:1, v/v), and vortex the content for ca 3 s leaving the organic phase milky.

7. Add a second portion (25 μL) of the catalytic medium, and shake the content for 5 s until the milky phase becomes clarified.

Table 2
GC-EI-SIM-MS operating conditions

GC Injector	Mode	Pulsed splitless (elevated head pressure from 110 to 220 kPa)
	Liner	Sky® 4 mm I.D. cyclo double taper inlet liner (Restek, P/N 23310)
	Temperature	220 °C
	Injection volume	1 μL
	Temperature	220 °C
GC oven	Initial temperature	75 °C
	Ramp	6 °C/min to 150 °C, 8 °C/min to 190 °C, 12 °C/min to 250 °C and at 20 °C/min to 300 °C hold for 3 min
	Run time	25 min
GC column	Capillary column	DB-XLB type, 30 m × 0.25 mm ID, 0.25 μm film thickness (Agilent, P/N 122-1232)
	Carrier gas	Helium
	Flow rate	1.2 mL/min
	Mode	Constant flow
	Outlet pressure	Vacuum
GC-MS transfer line	Temperature	250 °C
MS quadrupole	Ion source temperature	230 °C
	Full scan mode	m/z 50–800 Da
	SIM mode	Metabolite SIM m/z ions; *see* Table 1
Software	MSD ChemStation (version E.02, Agilent)	

8. Add 50 μL of the isooctane extraction medium, and mix for about 1–2 s.

9. Add of 25 μL of 1 M aqueous HCl and vortex briefly; if the phases are not well separated, a minicentrifuge may be a convenient option (*see* **Note 8**).

10. Aspirate 70–80 μL of the upper organic phase into a vial insert (150–200 μL volume).

11. Inject a sample extract aliquot (1 μL) by using a pulsed splitless injection into a GC injector, and start the GC-MS acquisition (*see* **Note 9**).

3.4 GC-MS Analysis

1. The instrument GC-MS conditions are summarized in Table 2.

2. First, analyze the standard mixtures to check the separation performance, retention times, the analyte peak shape, and

acquisition of the employed fragment ions in the obtained EI spectra (Table 1).

3. Using a single quadrupole MS analyzer, single-ion monitoring (SIM) mode is commonly used for quantification. Use the characteristic m/z ions (a quantifier and a qualifier) listed in Table 1. The SIM scanning should be arranged into convenient time sequence groups (windows) associating closely co-eluting analytes m/z ions thus enabling an appropriate SIM dwell time for each detected analyte [12].

4. Prepare an appropriate analysis sequence for a sample series consisting of repeated blank, standard, the QCs (*see* Subheading 3.2, **step 2**), calibration, and real sample extracts. Measure the blanks, QC samples, and standards regularly (at least every ten sample runs). Analyze samples in a random order to avoid systematic errors.

5. Change the GC injector liner after approximately 150 samples depending on the sample matrix. Before use, condition each new liner by running the following sequence: solvent blank, standard mixtures, the pooled QC sample extract (twice), and solvent blank (twice).

3.5 Data Analysis

1. Peak area for quantifier and qualifier ion of each metabolite is integrated. Their ratio is calculated to test for potential interferences.

2. The peak area of each quantifier is normalized by the peak area of the corresponding internal standard: amino acids and biogenic amines against homophenylalanine, compound No. 75; organic acids against 4-phenylbutyric acid, compound No. 37, Table 1 [12].

3. Use appropriate vendor data processing software for data calibration and metabolite quantification.

4. Check metabolite responses in the QC samples measured regularly throughout the whole sample set. If the analyte's relative response to the IS fluctuates with RSD >30%, then even a semi-quantitative measurement of such metabolite is difficult, and it should be excluded from the metabolomic study (*see* **Note 10**).

5. Once the metabolite levels have been determined and met, pre-defined acceptance criteria normalize appropriately the measured metabolite concentrations relative to creatinine. For urine recalculate the metabolite levels to creatinine or other suitable reference factors. Export the analytical data matrix into a Microsoft Excel® spreadsheet or other formats suitable for further chemometric analysis.

6. Use an appropriate statistical software to recognize differences among the studied metabolite sample sets. The calculation of

p-values by means of a *t*-test helps to determine significance of the obtained results. The data set can be conveniently examined graphically by means of box plots that display patterns of quantitative data and thus facilitate interpretation of observed metabolite changes in the studied organism.

4 Notes

1. For serum/plasma, (lipo)proteins must be precipitated with a suitable medium prior to the application of this protocol. The workup requires first an internal solution addition (step 2), which may be followed by the THP reduction of disulfide bonds (step 3). However, the protein precipitation must precede step 4 pH adjustment; it can be carried out with perchloric acid [13], trichloroacetic acid, or an organic solvent [15, 16]. Note that the selected precipitation conditions for plasma or serum may affect the obtained metabolite profile [15] and thus the described protocol must always be adapted and validated to a particular demand.

2. The 5% stock THP solution can be kept in a freezer for a year and the 0.5% working solution in a refrigerator for a month.

3. The 2,2,3,3,4,4,4-heptafluorobutyl chloroformate (HFBCF) reagent is a liquid with a boiling point 105–107 °C and density 1.6 g/cm^3 [14]. The reagent must be stored in tightly closed Teflon-lined cap glass vials at 4 °C and thus is stable for at least 24 months.

    WARNING! Manipulation with HFBCF must be performed in a well-ventilated area (fume hood).

4. A list of metabolites covered by the protocol is summarized in Table 1. For the analytical purposes, the compounds can be sorted into six groups, i.e., protein amino acids, non-protein amino acids, dicarboxylic acids, hydroxycarboxylic acids, organic (aromatic) acids, and metabolites sensitive to storage conditions. The last group comprised of oxoacids, lactate, 5-hydroxyindol- and indolacetate, 3- and 4-hydroxyphenylacetate, 4-hydroxymandelate, 4-hydroxyphenyllactate, kynurenate, vanillylmandelate, kynurenine, glycylproline, prolylhydroxyproline, and 3-methylcrotonylglycine; the stock and working solutions of this group were kept in freezer at −20 °C.

5. Urine like other important biological matrices is a metabolite-rich mixture. Be careful and always take into account properties of each metabolite of interest in the studied matrix. Check carefully the metabolite stability by sample measurement

within a convenient time period before making final conclusions from the measured data.

6. In metabolomics, normalization of samples is an important practice [17]. For urinary analysis, creatinine is the commonly used reference that indicates the urine concentration [18]. Urine of healthy women and men typically contains around 5–16 mmol/L of creatinine [12, 19]. If its concentration is highly increased (more than 3–4 times), the sample should be diluted with DI water before analysis maintaining thus the urine composition closer the average creatinine abundance.

7. The used HFBC reagent volume (10 μL, 60 μmol) is efficient for workup of urine volumes below 50 μL. In contrast with classical alkyl chloroformates, the corresponding heptafluorobutyryl alcohol is not necessary in the reaction medium. Nevertheless, a small 5% aliquot facilitates esterification of polycarboxylic acids such as citrate.

8. The acidification step substantially further decreases the pyridine catalyst content in the arising upper organic layer and thus contamination of the GC-MS system. Consequently, the liner change typically follows after 120–150 samples, less frequently than in earlier methods [20].

9. If the prepared organic sample extracts are not measured immediately, they can be stored in tightly closed Teflon-lined autosampler vials for up to 2 weeks at −20 °C. A slow degradation of some metabolites was observed, in particular kynurenate, tiglylglycine, 3-hydroxybutyrate, fumarate and prolylhydroxyproline, histidine, 1-methyl- and 3-methylhistidine, and isocitrate.

10. Note that demands on metabolomic analysis do not always conform to strict guidelines requested, for instance, by guidelines in drug analysis, and data showing a higher uncertainty may be useful in the study. Moreover, metabolite concentrations observed between two studied models rarely change by more than one order of magnitude, and thus narrower calibration ranges can be used throughout a metabolomic study with respect to the amount estimated in the pooled QC sample.

Acknowledgments

This work was supported by the Czech Science Foundation, project No. 17-22276S.

References

1. Andrews MA (1989) Capillary gas-chromatographic analysis of monosaccharides – improvements and comparisons using trifluoroacetylation and trimethylsilylation of sugar o-benzyl-oximes and o-methyl-oximes. Carbohydr Res 194:1–19. https://doi.org/10.1016/0008-6215(89)85001-3

2. Kostal V, Zahradnickova H, Simek P et al (2007) Multiple component system of sugars and polyols in the overwintering spruce bark beetle, Ips typographus. J Insect Physiol 53(6):580–586. https://doi.org/10.1016/j.jinphys.2007.02.009

3. Hill M, Parizek A, Kancheva R et al (2010) Steroid metabolome in plasma from the umbilical artery, umbilical vein, maternal cubital vein and in amniotic fluid in normal and preterm labor. J Steroid Biochem Mol Biol 121(3–5):594–610. https://doi.org/10.1016/j.jsbmb.2009.10.012

4. Rimnacova L, Husek P, Simek P (2014) A new method for immediate derivatization of hydroxyl groups by fluoroalkyl chloroformates and its application for the determination of sterols and tocopherols in human serum and amniotic fluid by gas chromatography-mass spectrometry. J Chromatogr A 1339:154–167. https://doi.org/10.1016/j.chroma.2014.03.007

5. Simek P, Heydova A, Jegorov A (1994) High-resolution capillary gas-chromatography and gas-chromatography mass-spectrometry of protein and nonprotein amino-acids, amino-alcohols, and hydroxycarboxylic acids as their tert-butyldimethylsilyl derivatives. J High Resoult Chromatogr 17(3):145–152

6. Kanani HH, Klapa MI (2007) Data correction strategy for metabolomics analysis using gas chromatography-mass spectrometry. Metab Eng 9(1):39–51. https://doi.org/10.1016/j.ymben.2006.08.001

7. Villas-Boas SG, Smart KF, Sivakumaran S, Lane GA (2011) Alkylation or silylation for analysis of amino and non-amino organic acid by GC-MS? Meta 1(1):3–20. https://doi.org/10.3390/metabo1010003

8. Husek P (1997) Urine organic acid profiling by capillary gas chromatography after a simple sample pretreatment. Clin Chem 43(10):1999–2001

9. Husek P, Simek P (2006) Alkyl chloroformates in sample derivatization strategies for GC analysis. Review on a decade use of the reagents as esterifying agents. Curr Pharm Anal 2(1):23–43. https://doi.org/10.2174/157341206775474007

10. Smart KF, Aggio RBM, Van Houtte JR et al (2010) Analytical platform for metabolome analysis of microbial cells using methyl chloroformate derivatization followed by gas chromatography-mass spectrometry. Nat Protoc 5(10):1709–1729. https://doi.org/10.1038/nprot.2010.108

11. Wachsmuth CJ, Hahn TA, Oefner PJ et al (2015) Enhanced metabolite profiling using a redesigned atmospheric pressure chemical ionization source for gas chromatography coupled to high-resolution time-of-flight mass spectrometry. Anal Bioanal Chem 407(22):6669–6680. https://doi.org/10.1007/s00216-015-8824-x

12. Husek P, Svagera Z, Hanzlikova D et al (2016) Profiling of urinary amino-carboxylic metabolites by in-situ heptafluorobutyl chloroformate mediated sample preparation and gas chromatography-mass spectrometry. J Chromatogr A 1443:211–232. https://doi.org/10.1016/j.chroma.2016.03.019

13. Simek P, Husek P, Zahradnickova H (2008) Gas chromatographic-mass spectrometric analysis of biomarkers related to folate and cobalamin status in human serum after dimercaptopropanesulfonate reduction and heptafluorobutyl chloroformate derivatization. Anal Chem 80(15):5776–5782. https://doi.org/10.1021/ac8003506

14. Simek P, Husek P, Zahradnickova H (2012) Heptafluorobutyl chloroformate-based sample preparation protocol for chiral and nonchiral amino acid analysis by gas chromatography. In: Alterman MA, Hunziker P (eds) Amino acid analysis: methods and protocols, Methods in molecular biology, vol 828, pp 137–152. https://doi.org/10.1007/978-1-61779-445-2_13

15. Husek P, Svagera Z, Hanzlikova D et al (2012) Survey of several methods deproteinizing human plasma before and within the chloroformate-mediated treatment of amino/carboxylic acids quantitated by gas chromatography. J Pharm Biomed Anal 67–68:159–162. https://doi.org/10.1016/j.jpba.2012.04.027

16. Svagera Z, Hanzlikova D, Simek P et al (2012) Study of disulfide reduction and alkyl chloroformate derivatization of plasma sulfur amino acids using gas chromatography-mass spectrometry. Anal Bioanal Chem 402(9):2953–2963. https://doi.org/10.1007/s00216-012-5727-y

17. Wu YM, Li L (2016) Sample normalization methods in quantitative metabolomics. J Chromatogr A 1430:80–95. https://doi.org/10.1016/j.chroma.2015.12.007

18. Bouatra S, Aziat F, Mandal R et al (2013) The human urine metabolome. PLoS One 8(9). https://doi.org/10.1371/journal.pone.0073076

19. Cimlova J, Kruzberska P, Svagera Z et al (2012) In situ derivatization-liquid liquid extraction as a sample preparation strategy for the determination of urinary biomarker prolyl-4-hydroxyproline by liquid chromatography-tandem mass spectrometry. J Mass Spectrom 47(3):294–302. https://doi.org/10.1002/jms.2952

20. Dettmer K, Stevens AP, Fagerer SR et al (2012) Amino acid analysis in physiological samples by GC-MS with propyl chloroformate derivatization and iTRAQ-LC-MS/MS. In: Alterman MA, Hunziker P (eds) Amino acid analysis: methods and protocols, Methods in molecular biology, vol 828, pp 165–181. https://doi.org/10.1007/978-1-61779-445-2_15

Sheathless Capillary Electrophoresis-Mass Spectrometry for the Profiling of Charged Metabolites in Biological Samples

Rawi Ramautar

Abstract

Capillary electrophoresis (CE) is well suited for the profiling of highly polar and charged metabolites as compounds are separated on the basis of their charge-to-size ratio. The protocol presented here is based on using a recently developed sheathless interfacing design, i.e., a porous tip interface, for coupling CE to electrospray ionization mass spectrometry (MS). It is demonstrated that sheathless CE-MS employing a bare fused-silica capillary at low-pH separation conditions can be used for the profiling of both cationic and anionic metabolites by only switching the MS detection and electrophoretic separation voltage polarity. The proposed sheathless CE-MS protocol allows efficient and sensitive profiles to be obtained for a broad array of charged metabolites, including amino acids, organic acids, nucleotides, and sugar phosphates, in various biological samples, such as urine and extracts of the glioblastoma cell line.

Key words Capillary electrophoresis, Mass spectrometry, Sheathless interface, Metabolomics, Biological samples, Cationic metabolites, Anionic metabolites

1 Introduction

A key aim of metabolomics is to obtain an answer to a specific research question, which may be of biological or clinical origin [1]. To achieve this goal, advanced analytical separation techniques are often used for the global profiling of (endogenous) metabolites in biological samples [2]. At present, the profiling of endogenous metabolites is generally performed with MS in combination with an online front-end chromatographic separation method [3]. Regardless of important developments in liquid chromatography column technology and methodology, the selective and efficient analysis of highly polar and charged compounds is still highly challenging. Capillary zone electrophoresis, referred to here as CE, separates compounds on the basis of differences in their intrinsic electrophoretic mobility, which is dependent on the charge and size of the analyte. Therefore, CE is highly suited for the analysis

Georgios A. Theodoridis et al. (eds.), *Metabolic Profiling: Methods and Protocols*, Methods in Molecular Biology, vol. 1738,
https://doi.org/10.1007/978-1-4939-7643-0_12, © Springer Science+Business Media, LLC, part of Springer Nature 2018

of polar and charged metabolites. Moreover, as the separation mechanism of CE is fundamentally different from chromatographic-based separation techniques, a complementary view on the composition of metabolites present in a given biological sample is provided.

Currently, the use of CE-MS in the field of metabolomics is relatively low as compared to other analytical techniques [4]. CE-MS is still considered a technically challenging approach by the overall scientific community, suffering from a relatively poor reproducibility and sensitivity. However, important to stress here is that CE-MS has been used for the global profiling of native peptides and endogenous metabolites in a clinical context for more than a decade now. For example, Mischak and coworkers have analyzed peptides in more than 20,000 human urine samples at different laboratories with an acceptable interlaboratory reproducibility [5, 6].

Soga and coworkers were the first to show the utility of CE-MS for the global profiling of metabolites in biological samples [7, 8]. CE is generally coupled to MS via a sheath-liquid interfacing technique [9, 10]; however, due to dilution of the capillary effluent by the sheath liquid, the detection sensitivity is intrinsically compromised.

Recently, it was demonstrated that the use of a sheathless interface significantly improved the detection coverage of metabolites present in various biological samples as compared to CE-MS employing a classical sheath-liquid interface [11–14]. The sheathless interface used was based on a porous tip emitter, which was invented by Moini [15], allowing the effective use of the intrinsically low-flow property of CE in combination with nano-ESI-MS.

In order to demonstrate the usefulness of sheathless CE-MS and to further expand the role of this method in metabolomics, a protocol is presented describing how this approach can be used for the analysis of charged metabolites in various biological samples, exemplified here for an extract of the human glioblastoma cell line and human urine. Some of the procedures outlined in this protocol has also been demonstrated in a visual manner recently [16]. Here, it is shown that by only switching the MS detection and electrophoretic separation voltage polarity, a single sheathless CE-MS method can be used for the profiling of cationic and a wide range of anionic metabolites using exactly the same capillary and separation conditions, thereby reducing analysis time for the global profiling of charged metabolites.

2   Materials

Prepare all solutions using ultrapure water (prepared by purifying deionized water to obtain a sensitivity of 18 MΩ-cm at 25 °C) and analytical-grade reagents.

2.1 Solutions and Samples for Analysis

1. *Background electrolyte (BGE) solution*: 10% (v/v) acetic acid, pH 2.2. Add 9.0 mL of water into a 10 mL glass vial, and add 1.0 mL of acetic acid to the water in a fume hood. Mix the solution thoroughly using a vortex. Store at 4 °C.

2. *Metabolite standard mixture*: dissolve 50 µL of a 50 µM cation standard mixture containing the basic twenty L-amino acids into 50 µL of water, and mix the solution thoroughly (*see* **Note 1**). Store at −80 °C when not in use. Dissolve 50 µL of a 50 µM anion standard mixture containing 17 anionic metabolites into 50 µL of water, and mix the solution thoroughly (*see* **Note 1**). Store at −80 °C when not in use. The anionic metabolites include (1) 2-naphthol-3,6-disulfonic acid; (2) D(+)2-phosphoglyceric acid; (3) D-ribose-5-phosphate; (4) D-glucose-1-phosphate; (5) D-glucose-6-phosphate; (6) D-fructose-6-phosphate; (7) inosine 5′-monophosphate; (8) guanosine 3′,5′-cyclic monophosphate; (9) guanosine 5′-mono-phosphate; (10) citric acid; (11) trimesic acid; (12) isocitric acid; (13) gluconic acid; (14) adenosine 3′,5′-cyclic monophosphate; (15) 2-hydroxybutyric acid; (16) b-diphosphopyridine nucleotide (NAD+); and (17) 3-hydroxybutyric acid.

2.2 Analytical Equipment

1. The protocol reported here can only be performed with a commercially available sheathless CE equipment, also known as CESI 8000 (Sciex, A98089). Dependent on the type of MS instrument, a dedicated nanospray source is required for hyphenating sheathless CE to MS, information which can be obtained from the vendor.

2. For the electrophoretic separations, commercially available fused-silica capillaries (dimensions, 30 µm ID × 90 cm total length) are used (Sciex, B07367).

3 Methods

The protocol described here for the use of sheathless CE-MS for metabolic profiling studies is for laboratory use only. Prior to using this protocol, consult all relevant material safety data sheets (MSDS). Please use all appropriate laboratory safety procedures, including safety glasses, lab coat, and gloves, when performing the experiments described in this protocol.

3.1 Setting Up the CE System

1. Place a new bare fused-silica cartridge with a porous tip emitter (30 µm ID × 90 cm total length) in the CE instrument.

2. Check for flow of liquid through the capillary by performing a forward rinse at 50 psi for 15 min using 100% methanol (*see* **Note 2**). Carry out also a rinse in the opposite direction at

50 psi for 5 min using 100% methanol to check the flow of liquid through the conductive capillary (*see* **Note 3**).

3. Rinse the separation capillary with water at 50 psi for 10 min by keeping the porous tip section, that is, the sprayer tip, in a 50 mL Falcon tube containing 5 mL of water.

4. Rinse the separation capillary with 0.1 M NaOH at 50 psi for 10 min, then by water at 50 psi for 10 min, and finally with BGE at 50 psi for 10 min.

3.2 Coupling Sheathless CE to MS

1. Remove the sprayer tip of the fused-silica cartridge from the water tube, and install it in the nanospray source adapter for coupling to the MS instrument (*see* **Note 4**). The ESI voltage is set to 0 during this step.

2. Ensure that the height of the BGE vials in the CE instrument matches the height of the sprayer tip.

3. Check for flow of liquid through the conductive capillary by rinsing with BGE at 50 psi for 5 min (*see* **Note 5**).

4. Rinse the separation capillary with BGE at 50 psi for 10 min in the forward direction (*see* **Note 2**).

5. Position the porous tip emitter at the entrance of the MS inlet at a distance of circa 2–3 mm. Apply a voltage of 30 kV using a ramp time of 1 min, and start acquiring MS data in the m/z range from 65 to 1000 m/z for metabolic profiling studies using first an ESI voltage of 0 (*see* **Note 6**).

6. Set the ESI voltage to 1000 V while continue recording data. Increase the ESI voltage with increments of 100 V until a constant background signal is observed.

7. Optimize the porous tip emitter position with respect to the center of the MS inlet by moving it in the x, y, or z-direction in order to see which position provides the maximal and most stable MS signal (*see* **Note 7**).

8. After determining the optimal ESI voltage, set the ESI voltage to 0, and decrease the CE voltage from 30 kV to 1 kV using a ramp time of 5 min (*see* **Note 8**).

9. Create on the basis of the optimized parameters a CE-MS method for the analysis of metabolite standards and biological samples.

3.3 Preparation of Extracts from the Human Glioblastoma Cell Line

1. After culturing a human glioblastoma cell line U-87 MG (ATCC HTB-14) in Dulbecco's Modified Eagle Medium (Invitrogen) supplemented with 5% fetal calf serum, 4 mM of L-glutamine, 25 mM of D-glucose and 1 mM of sodium pyruvate, and 100 µg/mL of penicillin/streptomycin (*see* **Note 9**), wash the adherent human U-87 MG glioblastoma cells with

   ice-cold saline solution (0.9%; w/v) to remove the remaining residues of culture medium.

2. Add ice-cold methanol/water solution (8/2, v/v) to quench the cellular metabolism.

3. Scrape the adherent cells using a 25 cm cell scraper.

4. Collect the methanol/water solution in a tube and snap-freeze in liquid nitrogen.

5. Add chloroform to the methanol/water fraction (final ratio 8/8/2, v/v/v), and centrifuge the sample for 10 min at 4 °C and $16,100 \times g$.

6. Collect the methanol/water layer and evaporate this fraction using a SpeedVac concentrator. Reconstitute the dried material in 50 µL water for analysis by sheathless CE-MS. When not in use, store the sample at −80 °C.

3.4 Preparation of Human Urine Samples

1. Collect human urine samples of healthy subjects, pool the samples, and store at −80 °C prior to usage.

2. Prior to sheathless CE-MS analysis, mix the pooled urine sample with the BGE (1:1, v/v), and centrifuge for 10 min at 4 °C and $16,100 \times g$.

3.5 Analysis of Metabolite Standards and Biological Samples

1. Add 20 µL of the anionic or cationic metabolite standard mixture into an empty 100 µL microvial (PCR vial) which fits into a CE vial, and put this vial in the inlet sample tray.

2. Rinse the separation capillary with BGE at 50 psi for 3 min followed by sample injection at 2.0 psi for 60 s (~20 nL corresponding to circa 3% of the capillary volume). Then perform BGE injection at 1.0 psi for 10 s.

3. Start MS data acquisition and apply a voltage of −30 kV (ramp time of 1.0 min) and a pressure of 0.5 psi at the inlet for anionic metabolic profiling or only +30 kV (ramp time of 1.0 min) for cationic metabolic profiling for 30 min (*see* **Note 10**). After the 30 min electrophoretic separation, stop MS data acquisition, and decrease the CE voltage to −1 or +1 kV using a ramp time of 5 min.

4. Between sample injections, rinse the capillary with water, 0.1 M sodium hydroxide, and BGE each at 30 psi for 3 min.

5. Evaluate the recorded data by determining the migration times and the signal intensity of the analyzed anionic and cationic metabolite mixtures. Check whether the anionic metabolite standards appear in the region between 10 and 28 min (Fig. 1). Check also whether three structurally related isomers, i.e., D-glucose-1-phosphate, D-glucose-6-phosphate, and D-fructose-6-phosphate, are partially separated (*see* **Note 11**).

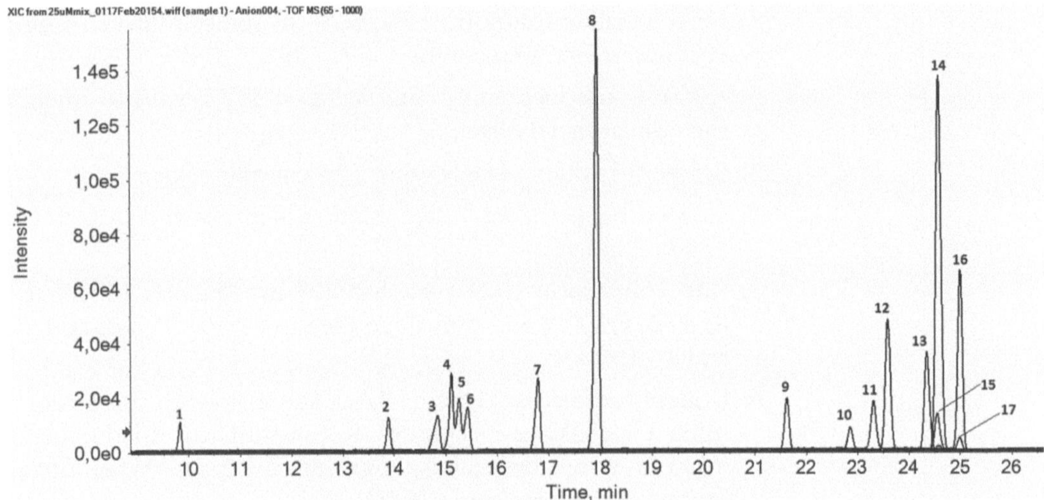

Fig. 1 Multiple extracted ion electropherograms obtained for the analysis of anionic metabolite standards metabolite (25 µM) with sheathless CE-MS in negative ion mode using a sheathless porous tip sprayer. Peaks: (1) 2-naphthol-3,6-disulfonic acid; (2) D(+)2-phosphoglyceric acid; (3) D-ribose-5-phosphate; (4) D-glucose-1-phosphate; (5) D-glucose-6-phosphate; (6) D-fructose-6-phosphate; (7) inosine 5′-monophosphate; (8) guanosine 3′,5′-cyclic monophosphate; (9) guanosine 5′-monophosphate; (10) citric acid; (11) trimesic acid; (12) isocitric acid; (13) gluconic acid; (14) adenosine 3′,5′-cyclic monophosphate; (15) 2-Hydroxybutyric acid; (16) b-Diphosphopyridine nucleotide (NAD+); (17) 3-Hydroxybutyric acid. Experimental conditions: BGE, 10% acetic acid (pH 2.2); separation voltage, −30 kV (+0.5 psi applied at the inlet of the CE capillary); sample injection, 2.0 psi for 60 s (Reproduced from ref. 13 with permission from the authors)

6.   Assess whether the cationic metabolite standards appear in the region between 8 and 22 min. Check whether isoleucine and leucine are migrating between 15 and 15.5 min, and determine if the resolution is circa 0.5 (Fig. 2).

7.   Use the procedures described in Subheading 3.3, **steps 1–6**, for anionic and cationic metabolic profiling of extracts of the glioblastoma cell line and human urine samples. A typical profile obtained for cationic metabolites in an extract from the glioblastoma cell line (cell density: ~20 cells/nL) by sheathless CE-MS is shown in Fig. 3.

8.   After analysis of the biological samples, analyze anionic and cationic metabolite standard mixtures to assess whether the performance of the sheathless CE-MS system is still adequate in terms of expected migration times, peak shapes, and detection sensitivity (*see* **Note 12**).

9. After the analyses or when not in use, rinse the capillary with water at 50 psi for 15 min, and store the inlet part of the capillary in a vial containing water and the porous section (outlet part) in a tube also containing water (*see* **Note 13**).

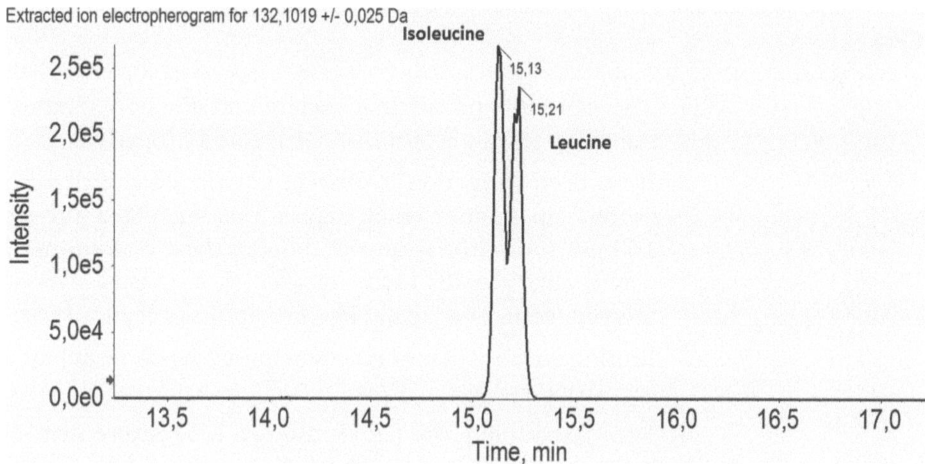

Fig. 2 Extracted ion electropherogram obtained for the analysis of isoleucine and leucine (25 μM) with sheathless CE-MS in positive ion mode. Experimental conditions: BGE, 10% acetic acid (pH 2.2); separation voltage, +30 kV; sample injection, 2.0 psi for 60 s

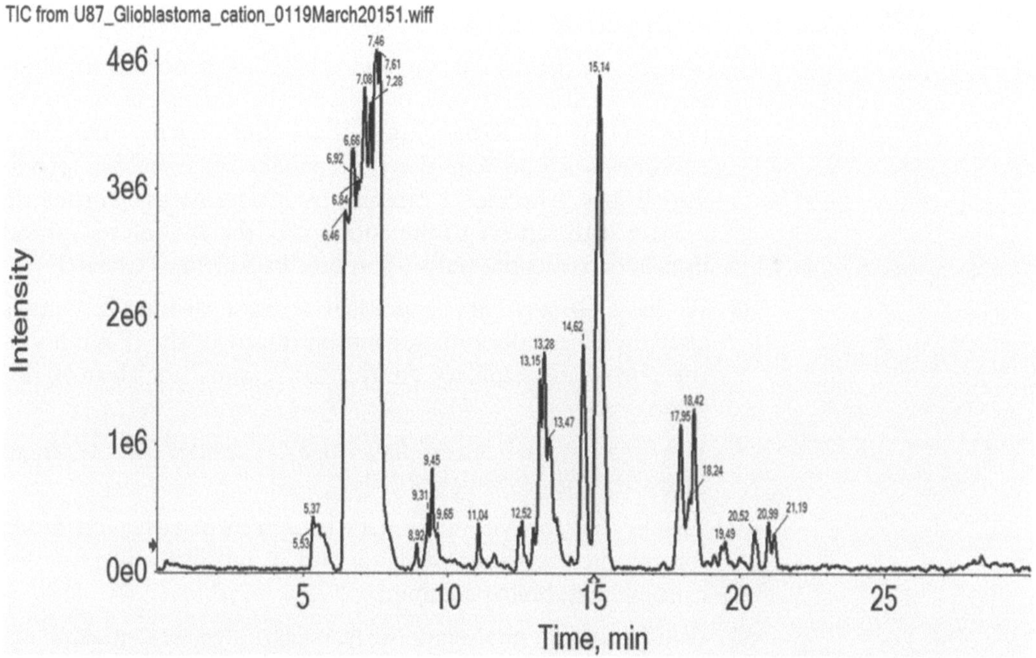

Fig. 3 Metabolic profile (total ion electropherogram) obtained for an extract of a glioblastoma cell line (cell density ~20 cells/nL) with sheathless CE-MS in positive ion mode. Experimental conditions: BGE, 10% acetic acid (pH 2.2); separation voltage, +30 kV; sample injection, 2.0 psi for 60 s (Reproduced from ref. 13 with permission from the authors)

4 Notes

1. The cationic and anionic metabolite standard mixtures are stable for at least 3 months when stored properly.

2. If no drop formation is observed at the end of the capillary porous tip emitter, then repeat this step at a pressure of 100 psi. If no drop is observed under these conditions, then a new capillary needs to be installed. For a visual overview of this procedure, we would like to refer to reference [16].

3. Rinsing with methanol is only required when installing a new capillary cartridge.

4. Prior to coupling the CE capillary to MS, ensure that the MS instrument has been calibrated and connected to the CE system.

5. During this rinsing step, a drop formation at the base of the ESI sprayer needle should be observed. If not, repeat the procedure at a pressure of 100 psi. Install a new capillary when no drop formation is observed under these conditions.

6. The mass spectrum should be void of signal as an electrospray voltage is not applied.

7. The sheathless CE-MS method is based on a porous tip emitter which allows the effective use of the intrinsically low-flow property of CE. Obtaining a stable ESI signal under these conditions is a prerequisite for reproducible metabolic profiling studies. Therefore, careful positioning of the porous tip emitter with respect to the entrance of the MS inlet is critical in order to obtain a stable/constant background signal.

8. We have observed that a gradual decrease of the CE voltage after the electrophoretic separation improves the durability of the porous tip capillary emitter for reasons not clear at this stage.

9. Use T75 cm^2 cell culture flasks at 37 °C under 5% CO_2 in an incubator for growing the cells.

10. Ensure that the MS instrument is used in negative ion mode for anionic metabolic profiling and in positive ion mode for cationic metabolic profiling.

11. Concerning the analysis of the isomers D-glucose-1-phosphate, D-glucose-6-phosphate, and D-fructose-6-phosphate (peaks 4, 5, and 6, respectively, in Fig. 1), the resolution between the first two peaks is circa 0.75 and of the last two peaks circa 0.50.

12. The analytical performance of the sheathless CE-MS method for metabolic profiling studies needs to be evaluated daily using metabolite standard mixtures. Under the same experi-

mental conditions, consistent migration times, i.e., variation below 3% for within-day ($n = 10$) and between-day ($n = 5$) using a 20 nL injection of a metabolite standard mixture (25 µM), peak areas (variation below 20%) and plate numbers (ranging between 50,000 and 400,000) are typically obtained. Detection limits in the nanomolar range are obtained for most metabolite standards under these conditions. In case these data are not obtained for the metabolite standards, the MS instrument needs to be tuned and recalibrated, or the porous tip capillary emitter needs to be changed.

13. When the sheathless CE-MS method is not in use, it is important to disconnect the separation capillary and to store the inlet side of the capillary in water and the outside submerged with the protective sleeve in a tube also containing water to prolong capillary lifetime. When proper rinsing and storage conditions are used in combination with a suitable sample pretreatment procedure, a single porous tip capillary can be used on average for 100 analyses.

Acknowledgment

Dr. Rawi Ramautar would like to acknowledge the financial support of the Veni and Vidi grant scheme of the Netherlands Organization for Scientific Research (NWO Veni 722.013.008 and Vidi 723.016.003).

References

1. Ramautar R, Berger R, van der Greef J et al (2013) Human metabolomics: strategies to understand biology. Curr Opin Chem Biol 17(5):841–846

2. Kuehnbaum NL, Britz-McKibbin P (2013) New advances in separation science for metabolomics: resolving chemical diversity in a postgenomic era. Chem Rev 113(4):2437–2468

3. Theodoridis GA, Gika HG, Want EJ et al (2012) Liquid chromatography-mass spectrometry based global metabolite profiling: a review. Anal Chim Acta 711:7–16

4. Ramautar R (2016) CE-MS in metabolomics: status quo and the way forward. Bioanalysis 8(5):371–374

5. Pejchinovski M, Hrnjez D, Ramirez-Torres A et al (2015) Capillary zone electrophoresis online coupled to mass spectrometry: a perspective application for clinical proteomics. Proteomics Clin Appl 9(5–6):453–468

6. Pontillo C, Filip S, Borras DM et al (2015) CE-MS-based proteomics in biomarker discovery and clinical application. Proteomics Clin Appl 9(3–4):322–334

7. Soga T, Ueno Y, Naraoka H et al (2002) Simultaneous determination of anionic intermediates for Bacillus subtilis metabolic pathways by capillary electrophoresis electrospray ionization mass spectrometry. Anal Chem 74(10):2233–2239

8. Soga T, Ohashi Y, Ueno Y et al (2003) Quantitative metabolome analysis using capillary electrophoresis mass spectrometry. J Proteome Res 2(5):488–494

9. Maxwell EJ, Chen DD (2008) Twenty years of interface development for capillary electrophoresis-electrospray ionization-mass spectrometry. Anal Chim Acta 627(1):25–33

10. Bonvin G, Schappler J, Rudaz S (2012) Capillary electrophoresis-electrospray ionization-mass spectrometry interfaces: fundamental concepts and technical developments. J Chromatogr A 1267:17–31

11. Bonvin G, Veuthey JL, Rudaz S et al (2012) Evaluation of a sheathless nanospray interface based on a porous tip sprayer for CE-ESI-MS coupling. Electrophoresis 33(4):552–562

12. Ramautar R, Busnel JM, Deelder AM et al (2012) Enhancing the coverage of the urinary metabolome by sheathless capillary electrophoresis-mass spectrometry. Anal Chem 84(2):885–892

13. Gulersonmez MC, Lock S, Hankemeier T et al (2016) Sheathless capillary electrophoresis-mass spectrometry for anionic metabolic profiling. Electrophoresis 37(7–8):1007–1014

14. Hirayama A, Tomita M, Soga T (2012) Sheathless capillary electrophoresis-mass spectrometry with a high-sensitivity porous sprayer for cationic metabolome analysis. Analyst 137(21):5026–5033

15. Moini M (2007) Simplifying CE-MS operation. 2. Interfacing low-flow separation techniques to mass spectrometry using a porous tip. Anal Chem 79(11):4241–4246

16. Zhang W, Gulersonmez MC, Hankemeier T et al (2016) Sheathless capillary electrophoresis-mass spectrometry for metabolic profiling of biological samples. J Vis Exp (116). doi:https://doi.org/10.3791/54535

Part III

Plant/Food Applications

Chapter 13

Two-Phase Extraction for Comprehensive Analysis of the Plant Metabolome by NMR

Jan Schripsema and Denise Dagnino

Abstract

Metabolomics is the area of research, which strives to obtain complete metabolic fingerprints, to detect differences between them, and to provide hypothesis to explain those differences [1]. But obtaining complete metabolic fingerprints is not an easy task. Metabolite extraction is a key step during this process, and much research has been devoted to finding the best solvent mixture to extract as much metabolites as possible.

Here a procedure is described for analysis of both polar and apolar metabolites using a two-phase extraction system. D_2O and $CDCl_3$ are the solvents of choice, and their major advantage is that, for the identification of the compounds, standard databases can be used because D_2O and $CDCl_3$ are the solvents most commonly used for pure compound NMR spectra. The procedure enables the absolute quantification of components via the addition of suitable internal standards. The extracts are also suitable for further analysis with other systems like LC-MS or GC-MS.

Key words Two-phase extraction, NMR, Metabolic fingerprints, Plants, Identification, Quantification

1   Introduction

Good planning is essential in any research project, even more so in metabolomic experiments due to the large numbers of samples typically analyzed. Every step should be made clear, from the question(s) to be answered to the methods and analytical techniques selected to achieve the goal. The whole procedure should be able to extract the maximum amount of information in a time- and cost-effective way. Once defined, the procedure should be strictly followed in order to allow the proper comparison of many samples [1].

NMR and MS, coupled to various chromatographic techniques, are today the main analytical methods used in metabolomics. Both methods have specific advantages and disadvantages and continue to develop and find useful applications.

Georgios A. Theodoridis et al. (eds.), *Metabolic Profiling: Methods and Protocols*, Methods in Molecular Biology, vol. 1738, https://doi.org/10.1007/978-1-4939-7643-0_13, © Springer Science+Business Media, LLC, part of Springer Nature 2018

Advantages of NMR- over MS-based metabolomics are the reproducibility of NMR spectra and the possibility of direct quantification [2]. Further a large dataset exists, available in the literature or specific data banks, containing spectra of pure compounds. Nevertheless, since the spectra vary according to the solvent used, direct comparison is only possible if the spectra to be compared have been obtained in the same solvent.

The extraction of the samples is a critical step in the metabolomics experiment, and much literature has been devoted to this subject, e.g. [3–5].

Most of the literature concerning NMR-based metabolomics uses mixtures of solvents in order to maximize the number of extracted metabolites [6–8]. But, when using this strategy, it is not possible to compare the spectra obtained directly with the readily available literature or databases.

In the present protocol, we focus on the extraction of plant material, taking into account the need to obtain complete metabolic fingerprints of the samples and the qualitative and quantitative determination of the compounds therein. We demonstrate the usefulness of a two-phase extraction system using water and chloroform. Both solvents are commonly used for pure compound NMR spectroscopic identification. Thus, the spectra obtained in metabolomic experiments can be readily compared with the available literature for the identification of the compounds therein. The spectra obtained with the two-phase method suggested are completely complementary, and very rarely does a compound appear in both phases. The separation in two phases diminishes superposition of signals (Fig. 1) and makes identification and quantification of the compounds in the extracts easier and more precise. It is a quick and extremely simple extraction procedure that can be carried out in targeted or untargeted approaches, and if necessary, the extracts can be further analyzed by LC-MS or GC-MS.

After obtaining the NMR spectra, the next steps in the metabolomic experiment involve the preprocessing of the obtained experimental results to make them suitable for the subsequent analysis that generally involves multivariate analysis. An important step in the preprocessing is the alignment of the signals in the different NMR spectra. To simplify the alignment, the process of binning or bucketing is generally used. Multivariate analysis will reveal the differences between sample groups, indicating the peaks or signals which are specifically increased or decreased in certain datasets.

In the final step of the experiments, these results should be interpreted and the compounds responsible for the signals identified.

In fact, one of the main difficulties in metabolomics is the discovery of the identity of the compounds detected. Peaks or signals are found which are specifically increased or decreased in certain

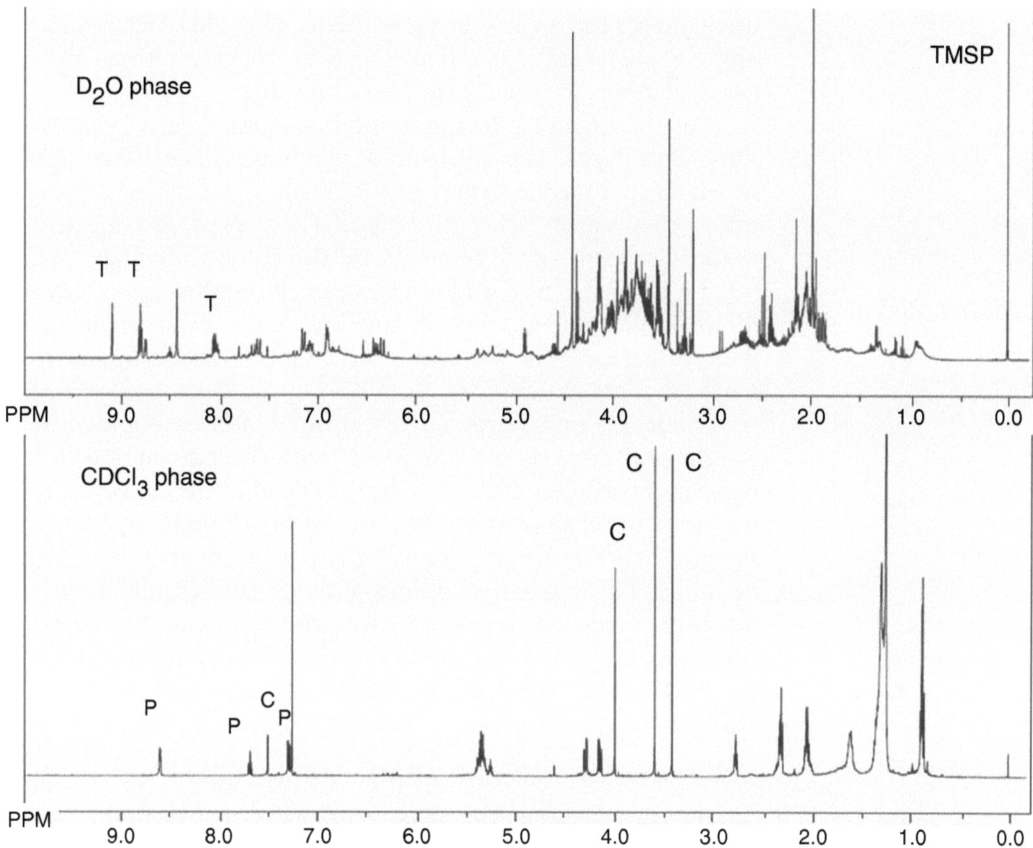

Fig. 1 ^{1}H NMR spectra from the (**a**) D$_2$O and (**b**) CDCl$_3$ extracts of coffee powder obtained with the two-phase extraction. In the CDCl$_3$ phase, signals from caffeine (marked with C) and pyridine (marked with P) are indicated. In the D$_2$O phase signals from TMSP and trigonelline (marked with T) are indicated

datasets, but linking those peaks to specific compounds needs a lot of care, and frequently additional experiments are required. For instance, in metabolomics based on LC-MS, major difficulties arise in the comparison of chromatograms (because of shifts in retention times) and/or spectra (reproducibility of fragmentation), when data obtained by different instruments or research groups are compared and when databases are consulted. In NMR spectroscopy the reproducibility is not such a problem, which makes the comparison of datasets easier.

2 Materials

2.1 Internal Standards

A known amount of internal standard solution is added to each of the phases so as to allow the quantification of the compounds extracted. For the aqueous phase, TMSP [3-(trimethylsilyl) proprionic-2,2,3,3-d4 acid, sodium salt, also reffered to as TSP] is used and for the organic phase pyridine. The TMSP spectrum has only one signal at 0.00 ppm making it suitable for both quantification and as a reference for the chemical shift.

For the organic phase pyridine is suggested since it provides some advantages. The spectrum of pyridine contains three signals of which two (at 8.61 ppm and 7.69 ppm) are clearly separated from other signals. The third signal (7.29 ppm) is close to the residual solvent signal and is more difficult to integrate. TMS, often used as chemical shift reference in chloroform, is not advised for quantification purposes, because it is more volatile and its signal is quite narrow. Furthermore its relaxation time is higher than most other compounds [9].

In metabolomics experiments, many signals are not identified, but the peak areas should be corrected to permit quantitative comparison between samples. First of all instead of the areas, the relative areas in relation to the area of the IS are used (area/area IS signal). This relative area should then be corrected for eventual deviations of the actual quantities of the IS added to the tube, the quantity of plant material extracted, and the quantity of solvent used to extract the plant material and of the liquid transferred to the NMR tube. This leads to the following multiplication factor (MF):

$$MF = \left(\text{Actual Weight IS sol.} / \text{Weight 100} \,.\, \mu\text{l IS sol.} \right)$$
$$\left(1000 \text{ mg of plant material} / \text{actual Quant.plant material} \right)$$
$$\left(\text{Quant.extraction solvent} / \text{Quant.extraction solvent transferred} \right).$$

This factor is the same for all signals in the NMR spectrum of a sample. In this way direct comparison between samples is possible.

If the compound of interest has well-resolved signals which can be integrated, the absolute quantity of the compound can be calculated using the following formula:

$$\text{Quant.IS in 100} \,.\, \mu\text{l IS sol. (mg)}$$
$$\left(\text{area compound signal} / \text{area IS signal} \right)$$
$$MF \;\; \left(\text{no.Hs IS signal} / \text{no.Hs compound signal} \right)$$
$$\left(\text{mol.weight compound} / \text{mol.weight IS} \right).$$

This formula provides the absolute quantity per gram of plant material. These calculations are illustrated in Table 1 for specific signals in the spectra shown in Fig. 1.

2.2 Preparation of the Internal Standard Solution (ISS)

Deuterated solvents should be at least 99.8% deuterated. The internal standards should be of analytical grade or higher. Weighing of standards and solvents should be carried out with maximum precision.

Table 1

Calculations with the NMR data shown in Fig. 1. For each spectrum the calculation of the absolute quantity of a specific compound is illustrated

	D$_2$O sample	CDCl$_3$ sample
	Trigonelline	Caffeine
	9.11 ppm	3.59 ppm
	IS 0.00 ppm	IS 8.61 ppm
IS solution	47.0 mg/100.0 g	151 mg/99.73 g
Actual weight IS sol.	110 mg	170 mg
Weight 100 µl IS sol.	110 mg	150 mg
Actual quant. plant material	105 mg	105 mg
Quant. extraction solvent	875 mg	1206 mg
Quant. extr. solvent transferred	524 mg	532 mg
Multiplication factor	15.90331	24.46831
Quant. IS in 100 µl IS sol. (mg)	0.0517 mg	0.227 mg
Area compound signal	62.54	252.29
Area IS signal	78.11	159.66
Number of Hs IS signal	9	2
Number of Hs compound signal	1	3
Molecular weight compound	137.14	194.19
Molecular weight IS	172.27	79.1
Abs. quant. (mg/g)	4.72	14.36

Internal standard solution 1 (ISS1): weigh 150 mg of pyridine (MW 79.10 g/mol) in an empty bottle and add 100 g CDCl$_3$.

Internal standard solution 2 (ISS2): weigh 50 mg of TMSP (MW 172.27 g/mol) in an empty bottle and add 100 g D$_2$O.

Internal standard solutions can be kept for up to 1 year at room temperature, if properly handled. Flasks should be kept tightly closed when not in use. In case of handling a large number of samples, distribute the prepared ISS in smaller bottles so as to minimize the time each flask is open.

3   Methods

3.1   *Extraction and Analysis*

The extraction described here has been used for dried plant material (freeze dried) and also for the analysis of dried bacteria and food stuffs such as teas, coffee, butter, and cheese. The procedure

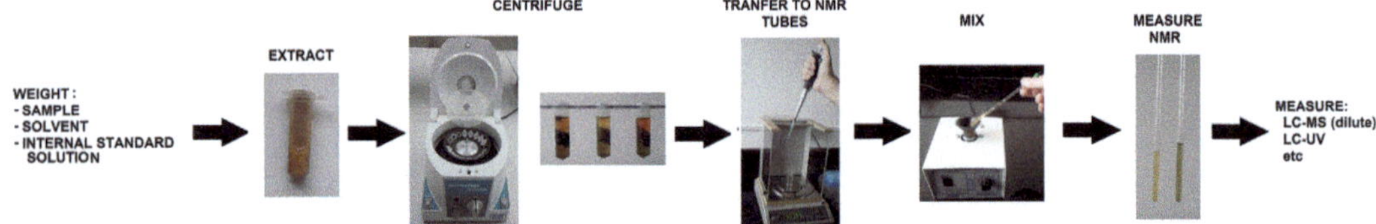

Fig. 2 Schematic representation of the two-phase extraction for comprehensive analysis of the plant metabolome by NMR

is schematically presented in Fig. 2. All pipetting should be carried out on a precision balance and the weights registered (*see* **Note 1**). When pipetting $CDCl_3$ with conventional pipettes, before the actual pipetting, aspire the liquid a few times to avoid dripping during pipetting.

1. Weigh 100 mg of dried plant material in a 2 ml Eppendorf tube (*see* **Notes 2** and **3**).

2. Add 0.80 ml of D_2O to the plant material, and make sure all material has been wetted (*see* **Note 4**).

3. Add 0.80 ml of $CDCl_3$ to the wetted plant material (*see* **Note 5**).

4. Thoroughly vortex the mixture.

5. Speed the extraction by ultrasonic mixing for 10 min.

6. Centrifuge at $13,800 \times g$ for 5 min.

7. Take out the Eppendorf tubes carefully from the centrifuge.

8. Add 0.10 ml of ISS1 to an NMR tube.

9. Take out 0.5 ml of the D_2O phase (upper phase) from the Eppendorf tube, and transfer it to the same NMR tube.

10. Seal the NMR tube, and vortex to mix the extract and the ISS1.

11. Take another NMR tube and add to it 0.10 ml of ISS2.

12. Take out 0.5 ml of the $CDCl_3$ phase (lower phase) with the pipette adjusted to 0.800 ml, and add to the same NMR tube (*see* **Notes 6–10**).

13. Seal the NMR tube, and vortex to mix the extract and the ISS2.

14. Acquire the ^1H NMR spectra (100–128 scans) of the extracts in both NMR tubes. For D_2O use presaturation of the residual solvent signal (*see* **Notes 11–14**).

4 Notes

1. Weighing increases the precision since pipetting is less precise, especially when chloroform is pipetted using conventional pipettes.

2. The proportion of sample to solvent can be changed depending on the concentration of the extracted compounds. The amount of sample chosen should preferably not saturate the solvents.

3. To determine the best proportion of sample to solvent volumes, re-extract the sample with the same procedure, and determine whether the extract contains around the predicted value. If not the proportion of sample to solvent can be adapted.

4. When the material is well wetted before the addition of $CDCl_3$, the extraction efficiency is increased.

5. $CDCl_3$ is added after the water to avoid evaporation.

6. Plant material stays between the two phases. When taking out the lower phase, the layer should be carefully traversed with the tip of the pipette. It is usually possible to displace the pellet formed between the phases by pushing it with the pipette tip and thus gain direct access to the lower phase.

7. When taking out the lower layer, chloroform vapors decrease the actual volume of the liquid. One should therefore adjust the pipette to 0.800 ml and take out as much of $CDCl_3$ as possible without taking along any of the aqueous phase.

8. When the other phase enters the pipette care should be taken not to transfer this to the NMR tube. The actual quantity which is transferred is determined by the weight of the liquid transferred to the NMR tube.

9. If a drop of the $CDCl_3$ extract falls on the balance plate, just allow it to evaporate (usually very quick) and register the weight after.

10. Minimum amount of volume in the NMR tube for a good quality measurement is 0.50–0.60 ml. If for any reason the volume is insufficient, record the weight of the extract and just add more deuterated solvent to complete the volume.

11. 100 or 128 scans are generally sufficient to obtain good-quality spectra, but the quantity can be increased (or decreased) according to the need.

12. After obtaining the spectra, they can be processed manually or automatically. This involves phasing and calibration. The peak shape should be verified to check if the sample was correctly shimmed. In the water samples, the TMSP signal can be

checked for symmetry. In the chloroform samples, the TMS or residual solvent signal can be used for this purpose. If not adequate the samples should be remeasured or discarded.

13. In the measurement of the aqueous extracts, presaturation of the residual solvent signal is used. In our experiments the Bruker program zgcppr was used. Typical parameters include the relaxation delay of 5.0 s. Acquisition of 64 K data points and 100 or 128 scans.

14. To analyze the extracts by mass spectrometry methods, it is necessary to dilute the extract in the deuterated solvents at least by a factor of 1000 with non-deuterated solvent. This will also eventually result in the exchange of H with D.

References

1. Schripsema J, Dagnino D (2015) Metabolomics. In: Hostettmann K, Stuppner H, Marston A, Chen S (eds) Handbook of chemical and biological plant analytical methods, 1st edn. Wiley, New York

2. Schripsema J (2010) Application of NMR in plant metabolomics: techniques, problems and prospects. Phytochem Anal 21:14–21

3. Deda O, Gika HG, Wilson IA, Theodoridis GA (2015) An overview of fecal preparation for global metabolic profiling. J Pharm Biomed Anal 113:137–150

4. Kim H-S, Park SJ, Hyun S-H et al (2011) Biochemical monitoring of black raspberry (*Rubus coreanus* Miquel) fruits according to maturation stage by ^{1}H NMR using multiple solvent systems. Food Res Int 44: 1977–1987

5. Heyman HM, Meyer JJM (2012) NMR-based metabolomics as a quality control tool for herbal products. S Afr J Bot 82:21–32

6. Kim HK, Verpoorte R (2010) Sample preparation for plant metabolomics. Phytochem Anal 21:4–13

7. Beltran A, Suarez M, Rodríguez MA et al (2012) Assessment of compatibility between extraction methods for NMR and LC/MS-based metabolomics. Anal Chem 84:5838–5844

8. Kim HK, Choi YH, Verpoorte R (2010) NMR-based metabolomic analysis of plants. Nat Protoc 5:536–549

9. Schripsema J (2008) Comprehensive analysis of polar and apolar constituents of butter and margarine by Nuclear Magnetic Resonance, reflecting quality and production processes. J Agric Food Chem 56:2547–2552

Chapter 14

NMR Spectroscopy Protocols for Food Metabolomics Applications

Evangelia Ralli, Maria Amargianitaki, Efi Manolopoulou, Maria Misiak, Georgios Markakis, Sofia Tachtalidou, Alexandra Kolesnikova, Photis Dais, and Apostolos Spyros

Abstract

NMR spectroscopy has become an indispensable tool for the metabolic profiling of foods and food products. In the present protocol, we report an analytical approach based on liquid-state NMR for the determination of polar and nonpolar metabolites in some common liquid (wine, spirits, juice) and solid (cheese, coffee, honey) foods. Although the diversity of foods precludes the use of a single protocol, with small modifications, the proposed methodologies can be adapted to a broader range of foodstuffs.

Key words NMR spectroscopy, Metabolite profiling, Food analysis, Authentication, Quality control

1 Introduction

High-resolution liquid-state NMR spectroscopy is a powerful analytical spectroscopic technique that can provide quantitative information on the chemical composition of liquid or solid but soluble mixtures of organic compounds. Compound identification and verification can be achieved either by comparison with literature data or public NMR spectroscopy databases. Even unknown compounds can be detected and identified by performing more complex 1D and 2D NMR experiments prior to quantification.

NMR spectroscopy has a long tradition in food analysis [1, 2] and in recent years has had a very significant contribution in the explosion of multivariate statistical analysis applications exploring the rich information inherent in the chemical profile (low MW organic compounds, usually secondary metabolites) of foods [3]. Together with mass spectrometry, NMR spectroscopy helped form a powerful pair of modern analytical techniques that became the workhorse and defined and shaped the field of modern food metabolomics [1, 4].

Georgios A. Theodoridis et al. (eds.), *Metabolic Profiling: Methods and Protocols*, Methods in Molecular Biology, vol. 1738, https://doi.org/10.1007/978-1-4939-7643-0_14, © Springer Science+Business Media, LLC, part of Springer Nature 2018

Foods can be liquid or solid, and their chemical composition diversity necessitates the use of a broad array of pre-analysis sample preparation procedures in order to extract useful metabolomic information from the analysis of NMR spectra. Liquid foods, containing a high concentration matrix of water or water/ethanol, represent the simpler analytical case scenario for NMR analysis (wine [5–7], spirits [8], juices [9]). These food samples can either be directly analyzed or, if increased sensitivity is necessary, the water/ethanol matrix can be easily removed via evaporation or freeze-drying (Subheading 3.1). Solid foods that are completely soluble in water or other solvents, such as honey, may also be directly analyzed by liquid-state NMR spectroscopy [10, 11]. In the case of solid insoluble foods, one can in principle resort to solid-state NMR methodologies utilizing high-resolution magic-angle spinning (HR-MAS) for food metabolomics applications, an approach that is gaining interest in recent years [12, 13]. However, for foods containing a significant amount of lipids, such as cheese [14–16] or coffee [17, 18], access to the water-soluble metabolome is limited by lipid interference, and vice versa, necessitating the application of more complicated extraction protocols for obtaining both the lipid and polar metabolite profiles (Subheading 3.2).

In the present protocol, we report two experimental protocols that can be used for the successful metabolomic analysis of several important foodstuffs. In the first protocol, simpler methodology proposed (Subheading 3.1), liquid foods can be studied directly (juice, alcoholic beverages, honey) or after water/ethanol removal (wine). In the second methodology proposed (Subheading 3.2), suitable extraction procedures are performed prior to the NMR spectroscopic determination of both the water-soluble metabolite and lipid profiles of the foods. We have applied methods in Subheading 3.2 to several types of hard and soft cheeses and several types of coffee (*see* Fig. 1) and cocoa grains, but it can be used with slight modifications for other complex and even whole foods.

2 Materials

2.1 Direct Analysis of Liquid or Water-Soluble Foods

1. Screw cap glass vials (2 mL).
2. A laboratory ultrasound sonicator.
3. A laboratory micro-centrifuge.
4. Analytical balance.
5. Variable volume pipette (100–1000 μL).
6. Ultrapure water.
7. 5 mm diameter glass NMR tubes (or appropriate size for the spectrometer probe used).

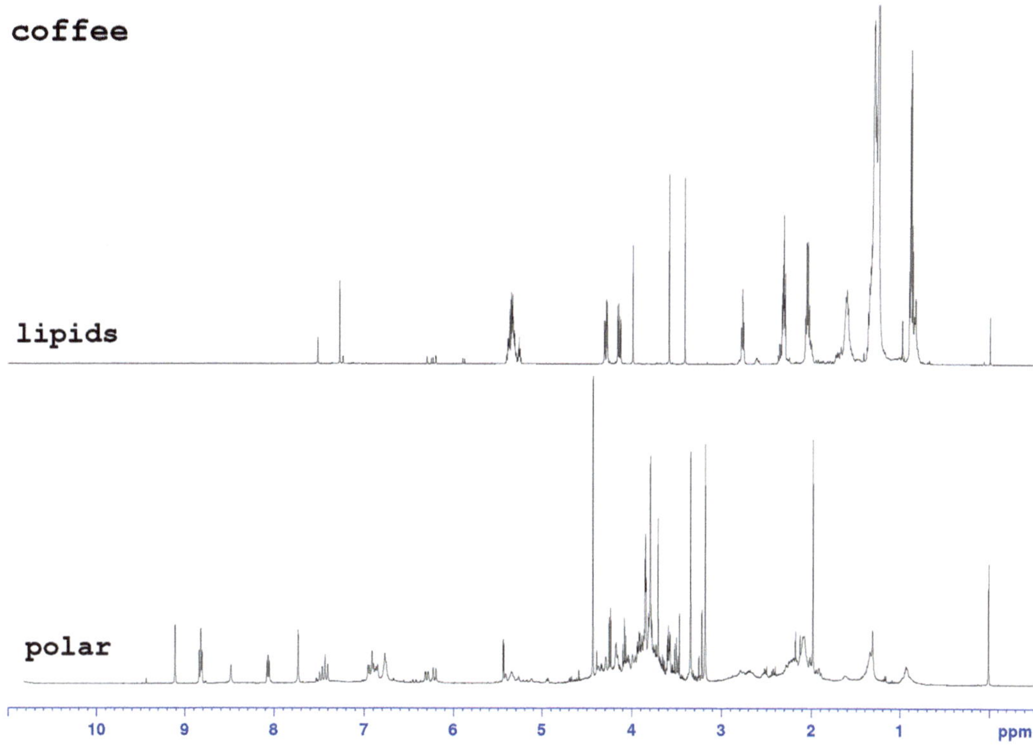

Fig. 1 ^1H NMR spectra of coffee obtained by applying method in Subheading 3.2: lipid profile in $CDCl_3$ (top) and polar metabolite profile in D_2O (bottom), both spectra obtained at a proton frequency of 500.13 MHz

8. Freeze dryer (Telstar Cryodos).

9. Deuterium oxide (D_2O) containing 0.05% TMSP (trimethyl-silyl propionic acid sodium salt, internal standard).

10. Oxalate buffer (pH = 4).

11. Phosphate buffer (pH = 7.4).

2.2 Analysis of Solid Foods

1. Screw cap glass vials (4 mL).

2. Eppendorf vials (1.5 mL).

3. A laboratory ultrasound sonicator.

4. A laboratory micro-centrifuge.

5. Analytical balance.

6. Variable volume pipette (100–1000 μL).

7. Pestle and mortar.

8. Liquid nitrogen (0.5 L).

9. Ultrapure water.

10. Glass flasks (50 or 100 mL).

11. 5 mm diameter glass NMR tubes (or appropriate size for the spectrometer probe used).

12. Freeze dryer (Telstar Cryodos).

13. D_2O containing 0.05% TMSP (trimethyl-silyl propionic acid sodium salt-d_4, internal standard).

14. $CDCl_3$ containing 0.03% TMS (tetramethylsilane, internal standard).

15. NMR spectrometer with high-resolution liquid-state probe.

2.3 Buffer Preparation

Oxalate buffer (pH = 4): transfer 0.0595 g of oxalic acid and 0.1795 g of sodium oxalate in a 20 mL volumetric flask, dissolve in D_2O (99.9 atom % D), and leave in ultrasound sonicator for 15 min.

Phosphate buffer (pH = 7.4): prepare phosphate buffer (pH 7.4) by weighing 0.721 g Na_2HPO_4, 0.131 g NaH_2PO_4, 1 mM TSP (0.025 g), and 3 mM NaN_3 (0.542 g) into a 25 mL volumetric flask. Add 5 mL of D_2O and fill up to 25 mL with water. Shake thoroughly, and leave in a sonicator at 40 °C, interspersed by shaking the flask, until the salts are dissolved.

3   Methods

3.1 Direct Analysis of Liquid or Water-Soluble Foods

3.1.1 Analysis of Wine

1. Store 2 mL of sample in a 4 mL screw cap vial at −18 °C for 12 h.

2. Lyophilize for at least 12 h.

3. Add 400 μL D_2O containing 0.05% TMSP and 200 μL oxalate buffer.

4. Centrifuge in an Eppendorf vial at 13, 1 48 × g for 10 min.

5. Transfer the supernatant into a 5 mm NMR tube.

6. Run NMR experiment Protocol A.

3.1.2 Analysis of Alcoholic Beverages (Spirits)

1. Transfer 600 μL of sample to a 5 mm NMR sample tube.

2. Add 100 μL of D_2O containing 0.05% TMSP.

3. Vortex for 1 min.

4. Run NMR experiment Protocol A.

3.1.3 Analysis of Honey

1. Weigh ~150 mg of sample in a 4 mL screw cap vial.

2. Add 600 μL of D_2O containing 0.05% TMSP.

3. Vortex for 1 min and transfer into a 5 mm NMR sample tube.

4. Allow to equilibrate for 2–3 days at room temperature in the dark (*see* **Note 1**).

5. Run NMR experiment Protocol B.

3.1.4 Analysis of Juice

1. Transfer 600 µL of juice to a 1.5 mL Eppendorf vial.
2. Centrifuge at 13,148 × g for 10 min.
3. Transfer 300 µL of the supernatant to a 5 mm NMR sample tube.
4. Add 200 µL of phosphate buffer and 100 µL of D_2O containing 0.05% TMSP.
6. Vortex for 1 min.
7. Run NMR experiment Protocol B.

3.2 Analysis of Solid Foods

3.2.1 Sample Pretreatment

1. Store 5 g of sample at −18 °C for 24 h.
2. Cut frozen sample to small pieces and weigh.
3. Place sample in 100 mL glass flask, connect to freeze dryer, and freeze-dry for 16 h.
4. Weigh samples again to calculate the amount of moisture.
5. Grind sample with a pestle and mortar under liquid nitrogen (*see* **Note 2**).

3.2.2 Polar Metabolite Extraction

6. Weigh 3× 0.30 g of ground sample into three Eppendorf vials (total 0.90 g).
7. Add 1 mL of ultrapure water in each vial by a variable volume pipette 100–1000 µL.
8. Seal vials with laboratory film, place into sonicator bath for 30 min, and centrifuge at 13,148 × g for 10 min (*see* **Note 3**).
9. Carefully remove the aqueous phase via pipette, and transfer into a tarred 50 mL glass flask.
10. Re-extract the remaining pellet two more times (repeat **steps 7–9**).
11. After the extraction in triplicate, store the three Eppendorf vials at −18 °C (to be used for the extraction of lipids in **step 17**) (*see* **Note 4**).
12. Freeze at −18 °C the glass flask with collected extracts and freeze-dry for 16 h.
13. After freeze-drying, weigh the precipitates, and transfer approx. half of the precipitate into a 4 mL screw cap glass vial.
14. Add 700 µL D_2O containing 0.05% TMSP via a variable volume pipette in the glass vial, and place into ultrasound sonicator bath for 30 min (*see* **Note 5**).
15. Filter the polar extracts through glass wool tightly packed into a Pasteur pipette, directly into a 5 mm NMR tube.
16. Run NMR experiment Protocol B.

17. Add 1 mL chloroform in each Eppendorf vial obtained in **step 11**.

18. Seal the three Eppendorfs with laboratory film, place into ultrasound bath for 30 min, and centrifuge at 10,000 rpm ($6,708 \times g$) for 10 min.

19. Carefully remove the liquid phase and transfer into a glass flask of 50 mL.

20. Re-extract the remaining pellet two more times with chloroform (repeat **steps 17–19**).

21. Evaporate the chloroform from the glass flask in a rotary evaporator.

22. Add 700 μL $CDCl_3$ containing 0.03% TMS (by the use of variable volume pipette 100–1000 μL) in the glass flask with the dried extracts, and place flask into ultrasound bath for 5 min.

23. Filter extracts carefully through glass wool directly into a 5 mm NMR tube.

24. Run NMR experiment Protocol C.

3.3 NMR Spectroscopy Experimental Protocols

3.3.1 Protocol A. ^{1}H NMR Spectroscopy WET Experiment

Set the probe temperature to 298 K, and wait (5–10 min) until the sample temperature is equilibrated. Lock, tune, and shim the sample according to standard NMR spectrometer procedures (*see* **Note 6**).

1. Obtain a ^{1}H NMR spectrum (Bruker, zg30); integrate and save the spectral regions containing the water and ethanol peaks to be suppressed per automation program directions (*see* **Note** 7).

2. Load a standard WET solvent multisuppression (Bruker/WET) pulse program with default spectrometer parameters.

3. Record a WET multisuppressed ^{1}H NMR spectrum with parameters SW = 20 ppm, NS = 256 scans, DS = 8 dummy scans, AQ = 3.3 s, and D1 = 1 s.

4. Perform Fourier transformation, phase correction, and baseline correction according to standard NMR spectrometer (or processing software) procedures.

3.3.2 Protocol B. ^{1}H NMR Spectroscopy Water Presaturation Experiment

1. Obtain a ^{1}H NMR spectrum (Bruker, zg30), and record the exact frequency of the residual water proton peak.

2. Load a standard solvent presaturation (Bruker, zgpr) pulse program with default spectrometer parameters, and set the frequency of the residual water signal exactly on resonance.

3. Record a water-suppressed ^{1}H NMR spectrum with parameters SW = 12 ppm, NS = 256 scans, DS = 4 dummy scans, AQ = 5.45 s, and D1 = 1 s.

4. Perform Fourier transformation, phase correction, and baseline correction according to standard NMR spectrometer (or processing software) procedures.

3.3.3 Protocol C. ^{1}H NMR Spectroscopy Standard Experiment

1. Load a standard proton NMR experiment (Bruker, zg30).

2. Record a ^{1}H NMR spectrum with parameters SW = 12 ppm, TD = 64 K, NS = 256 scans, DS = 4 dummy scans, AQ = 3.3 s, and D1 = 1 s.

3. Perform Fourier transformation, phase correction, and baseline correction according to standard NMR spectrometer (or processing software) procedures.

3.4 NMR Data Analysis

3.4.1 Metabolite Profiling

The use of an internal standard in the form of TMSP allows the quantitative determination of polar metabolites in the food samples from the integration of proton signals in the ^{1}H NMR spectra obtained. The ^{1}H NMR spectra of the nonpolar fraction of solid foods (such as cheese, coffee, cocoa) can be used to quantify the fatty acid profile of the lipid fraction of these foods. In either case, identification of metabolites can be achieved through spiking experiments (when the metabolite is expected or suspected) and verified through 2D NMR spectroscopy. Other means of metabolite identification include the use of publicly available NMR spectral databases, such as FoodDB, HMDB, BMRB, etc. Multivariate statistical analysis methods can be applied directly on the quantitative NMR metabolite profiles, in order to develop models for studying food authentication, quality control, cultivar, pedoclimatic effects, etc.

3.4.2 Metabolite Fingerprinting

For metabolite fingerprinting applications, the whole NMR spectrum is used for statistical analysis and development of multivariate metabolomic models. The first step is to convert the NMR spectra into ASCII files and/or use suitable software (AMIX) to segment the NMR spectra into buckets, for variable reduction purposes. The size (width) of the buckets is user-defined, usually between 0.005 and 0.01 ppm for ^{1}H NMR data, and the spectra are normalized to the total sum of integrals of all buckets. Regions of the spectra that do not contain any signal or contain solvent signal (e.g., water or ethanol) can be omitted from the bucketing procedure. If the NMR spectra need alignment, due to pH, temperature, or other sample variations, this should be performed using specialized software, before the bucketing procedure [19].

In the data matrix obtained, the different food samples make up the rows, while the buckets make up the matrix columns, respectively. The dataset can be subjected to multivariate statistical analysis using a variety of commercial and academic/public software packages, including STATISTICA, SIMCA, R, etc. Analysis of variance (ANOVA) and unsupervised principal component analysis (PCA) should be undertaken initially, in order to examine the dataset for potential outliers and remove them from subsequent analysis.

PCA can also be used for the observation of possible trends and relationships between the different categories/classes of food samples. In order to enhance the separation among the different classes of samples, supervised methods such as partial least squares discriminant analysis (PLS-DA) or orthogonal partial least squares discriminant analysis (OPLS-DA) can be used, using class membership as a dependent variable. PLS-DA and OPLS-DA models also offer the advantage of producing robust models that can be validated externally using test samples (food samples that have not been used for model development) and then used for class prediction of unknown samples. In all metabolomics models described, the loading plots obtained by various statistical methods indicate compound/spectral regions characteristic for particular traits of the food samples that are strong contributors to the obtained class separation/classification.

4   Notes

1. Full equilibration in water of the sugar molecules is necessary to avoid spectral differences due to nonequilibrium forms between samples.

2. For samples already in the ground form (coffee, cocoa), this step is omitted.

3. Three phases are formed in the Eppendorf vial, a white solid lipid phase at the bottom, an aqueous phase in the middle, and a thin nonpolar liquid phase at the top.

4. If lipid metabolite profiling is not required, the Eppendorf vials may be discarded.

5. For ground coffee/cocoa, an extra centrifugation of the sample solution (10 min) may be performed after **step 14**.

6. It is of utmost importance that all NMR experiments in a metabolomic dataset should be conducted under exactly the same conditions, i.e., NMR sample volume, pH, temperature, and pulse sequence parameters (receiver gain, number of scans, relaxation delay, etc.).

7. The WET experiment is included in standard Bruker spectrometer automation routines.

References

1. Spyros A, Dais P (2012) NMR spectroscopy in food analysis. RSC Food Analysis Monographs. Cambridge

2. Uryupin AB, Peregudov AS (2013) Application of NMR techniques to the determination of the composition of tobacco, coffee, and tea products. J Anal Chem 68(12):1021–1032. https://doi.org/10.1134/S1061934813120125

3. Monakhova YB, Kuballa T, Lachenmeier DW (2013) Chemometric methods in NMR spectroscopic analysis of food products. J Anal Chem 68(9):755–766. https://doi.org/10.1134/S1061934813090098

4. Spyros A (2016) Application of NMR in food analysis. In: Ramesh V (ed) Specialist periodical reports: nuclear magnetic resonance, vol 45. The Royal Society of Chemistry, Cambridge, pp 269–307. https://doi.org/10.1039/9781782624103-00269

5. Košir IJ, Kidrič J (2001) Identification of amino acids in wines by one-and two-dimensional nuclear magnetic resonance spectroscopy. J Agric Food Chem 49(1):50–56

6. Lee JE, Hwang GS, Van Den Berg F et al (2009) Evidence of vintage effects on grape wines using 1H NMR-based metabolomic study. Anal Chim Acta 648(1):71–76

7. Son HS, Ki MK, Van Den Berg F et al (2008) ¹H nuclear magnetic resonance-based metabolomic characterization of wines by grape varieties and production areas. J Agric Food Chem 56(17):8007–8016

8. Fotakis C, Kokkotou K, Zoumpoulakis P et al (2013) NMR metabolite fingerprinting in grape derived products: an overview. Food Res Int 54(1):1184–1194. https://doi.org/10.1016/j.foodres.2013.03.032

9. Belton PS, Colquhoun IJ, Kemsley EK et al (1998) Application of chemometrics to the ¹H NMR spectra of apple juices: discrimination between apple varieties. Food Chem 61(1–2):207–213

10. Spiteri M, Jamin E, Thomas F et al (2015) Fast and global authenticity screening of honey using ¹H-NMR profiling. Food Chem 189:60–66. https://doi.org/10.1016/j.foodchem.2014.11.099

11. Kazalaki A, Misiak M, Spyros A et al (2015) Identification and quantitative determination of carbohydrate molecules in Greek honey by employing ¹³C NMR spectroscopy. Anal Methods 7(14):5962–5972. https://doi.org/10.1039/c5ay01243k

12. Valentini M, Ritota M, Cafiero C et al (2011) The HRMAS-NMR tool in foodstuff characterisation. Magn Reson Chem 49(Suppl 1):S121–S125. https://doi.org/10.1002/mrc.2826

13. Ritota M, Casciani L, Failla S et al (2012) HRMAS-NMR spectroscopy and multivariate analysis meat characterisation. Meat Sci 92(4):754–761. https://doi.org/10.1016/j.meatsci.2012.06.034

14. Gianferri R, Maioli M, Delfini M et al (2007) A low-resolution and high-resolution nuclear magnetic resonance integrated approach to investigate the physical structure and metabolic profile of Mozzarella di Bufala Campana cheese. Int Dairy J 17(2):167–176

15. Lamanna R, Piscioneri I, Romanelli V et al (2008) A preliminary study of soft cheese degradation in different packaging conditions by ¹H-NMR. Magn Reson Chem 46(9):828–831

16. Schievano E, Pasini G, Cozzi G et al (2008) Identification of the production chain of Asiago d'Allevo cheese by nuclear magnetic resonance spectroscopy and principal component analysis. J Agric Food Chem 56(16):7208–7214

17. Bosco M, Toffanin R, De Palo D et al (1999) High-resolution ¹H NMR investigation of coffee. J Sci Food Agric 79(6):869–878

18. Consonni R, Cagliani LR, Cogliati C (2012) NMR based geographical characterization of roasted coffee. Talanta 88:420–426

19. Savorani F, Tomasi G, Engelsen SB (2010) Icoshift: a versatile tool for the rapid alignment of ¹D NMR spectra. J Magn Reson 202(2):190–202

Chapter 15

Direct Injection Analysis of Fruit VOCs by PTR-ToF-MS: The Apple Case Study

Brian Farneti

Abstract

The instrumental characterization of volatile organic compounds (VOCs) is essential to have a precise, reliable, and reproducible estimation of food aroma and, therefore, of the overall product quality. In this report, we introduce four analytical approaches based on PTR-MS (proton transfer reaction-mass spectrometry) technology suitable to fully investigate the complexity of apple aroma. In our opinion, these proposed methodologies can be applied, with slight modification, to every kind of fruit for destructive and nondestructive rapid VOC fingerprinting.

Key words PTR-ToF-MS, Profiling, VOCs, Aroma, Autosampler, Artificial chewing

1 Introduction

Studies focused on food quality improvement have to consider the complexity of flavor profile, its evolution in time, and the interaction with consumers. Volatile organic compounds (VOCs) are produced and released in most stages of the food production chain, "from farm to fork"; therefore they play a relevant role in agro-industrial processes.

The monitoring of VOCs produced by fruits and vegetables needs analytical techniques that are capable of dealing with challenging issues: (1) the need for separating and quantifying VOCs in complex gas mixtures; (2) the need to detect concentrations that may span a large range, from trace levels to parts per million; and (3) the need to track concentrations that rapidly change over time [1]. Because of these experimental constraints, the ideal methodology for VOC monitoring should be highly selective, with high sensitivity and dynamic range and with high time resolution.

Non-chromatographic techniques, based on direct injection mass spectrometric VOC assessment, are receiving great interest mainly (1) because of their capacity to carry out rapid, high-throughput measurement of large sample sets without affecting

Georgios A. Theodoridis et al. (eds.), *Metabolic Profiling: Methods and Protocols*, Methods in Molecular Biology, vol. 1738, https://doi.org/10.1007/978-1-4939-7643-0_15, © Springer Science+Business Media, LLC, part of Springer Nature 2018

samples and without interfering with the VOC production process and (2) because of the possibility of rapid process monitoring [1]. VOC monitoring for the abovementioned applications not only relies on high time resolution but also takes advantage of the high sensitivity provided by most direct injection mass spectrometry (DIMS) technologies. Rapid monitoring does not allow for pre-concentration, and the compounds of interest may well be present in the gas phase in the sub-ppt range nevertheless inducing a relevant effect.

Besides its technological performances (e.g., sensitivity and selectivity), advanced DIMS is also increasingly being used because of its stability since the mass/charge ratio does not vary with the experimental conditions. However, the greatest difficulty arising in DIMS, due to the lack of chromatographic separation, is the need to identify the compounds that generate the observed peaks, since the latter can be the results of overlapping signals from the mix of different VOCs present in the sample. A number of real-time mass spectrometric techniques are available for food VOC analysis, including MS-e-noses, atmospheric pressure chemical ionization mass spectrometry (APCI-MS), selected ion flow tube-MS (SIFT-MS), and proton transfer reaction-mass spectrometry (PTR-MS).

In the present protocol, we reported four analytical approaches based on PTR-MS technology suitable to fully investigate the complexity of apple aroma. In our opinion, these proposed methodologies can be applied, with slight modification, to every kind of fruit or vegetable.

1.1 PTR-MS Technology

PTR-MS is an emerging technique, developed in the mid-1990, that has already found a wide range of applications in ecological and environmental monitoring, agriculture and food sciences, and medical diagnostics [2, 3]. PTR-MS is a form of soft chemical ionization mass spectrometry based on proton transfer from a protonated reagent, most commonly H_3O^+. The fundamental ionization process can be written as

$$H_3O^+ + R \rightarrow RH^+ + H_2O$$

Protonated water (H_3O^+) interacts with the trace gas molecule (R). During this interaction, a proton transfers from the hydronium to the trace gas molecule, which leads to a protonated and therefore an ionized molecule (RH^+) and a neutral water molecule (H_2O). This proton transfer reaction is energetically favorable for all VOCs with a proton affinity higher than that of water (691 kJ/mol). This criterion excludes the major components of air such as N_2, O_2, and CO_2 (consequently not interfering with the measurement) but includes many trace gases including most volatile organic compounds (VOCs).

The main constituents of a PTR-MS apparatus are the ion source, the reaction region (drift tube), the mass analyzer, and the ion detector.

In most apparatus, the H_3O^+ primary ion beam is produced by a hollow cathode ion source. A promising development of the hollow cathode ion source is the use of primary parent ions other than H_3O^+ to allow the detection of compounds with proton affinities larger than that of H_2O. This possibility was already investigated by Yeretzian, Jordan, and Lindinger [4] using NH_4^+ ions to bracket compounds for coffee aroma analysis. More recently, other ions have been studied (e.g., O_2^+ or NO^+), leading to the development of a marketed commercial system allowing for rapid switching between different ions: O_2^+, NO^+, Kr^+, and Xe^+. Selected ion flow technique (SIFT) provides this unique advantage as well, but the better selection of the parent ions is paid by a lower sensitivity [5].

The second part of a PTR-MS apparatus is the drift region, where the parent ions are driven by an electric field and eventually interact with the VOC to be detected. The process is controlled by drift-tube temperature, pressure, and electric potential. Decreasing the drift-tube volume increases the maximum obtainable time resolution increasing reaction efficiency (higher sensitivity).

The last part of a PTR-MS system is the mass analyzer. Linear quadrupoles were used in most instruments. They are robust and relatively cheap, but have time resolution of the order of seconds, and typically provide only unit mass resolution. An alternative to overcome these limitations of the quadrupole mass analyzer is the coupling of PTR with a time-of-flight (ToF) mass analyzer. The great advantage of ToF analyzers, built upon the observation that at the same kinetic energy, heavier ions fly more slowly than lighter ones, is the enhanced analytical information provided. In fact, they may reach a resolution up to 7000 $(m/\Delta m)$ for commercial PTR-MS instruments, thus allowing the separation of many isobaric compounds and the simultaneous monitoring of multiple peaks at the same nominal mass. Moreover, if proper mass calibration is applied to ToF spectra, it is also possible to identify the compound sum formula. The whole spectrum is acquired in a split second, while a scan with a quadrupole can take a minute or more to get the same sensitivity.

1.2 Nondestructive Analysis of Intact Fruit	Headspace VOC fingerprint by PTR-MS provides a potential tool for discriminating fruit and vegetable not only based on genetic differences but also based on origin and maturity stages. The nondestructive application of PTR-MS may be not only important for cultivar characterization but also in product origin separation and physiological experiments related to fruit ripening and quality.
1.3 Destructive Analysis of Fresh Fruit Tissue	The correlation between VOC emitted by an intact fruit and its internal content is still commonly accepted. Indeed, most of the investigations carried out on VOC composition in fruit, and related

with physiological and genetic studies, are based on static headspace analysis of intact fruit.

However, the blend of apple aroma compounds is subjected to important modifications of severalfold change due to fruit cutting. Therefore, the destructive analytical assessment of fresh fruit tissue is essential in particular for studies related to fruit quality.

1.4 Dynamic Analysis by In Vitro Mastication

The flavor of a product is not a stable trait, but it can rapidly and drastically change during time, for instance, during fruit consumption. Differences in VOC release behaviors may influence the human aroma perception during food consumption since VOCs are released from the matrix and then transported to mouth and nose receptors.

In order to describe the release kinetics of VOCs while the food matrix is being chewed, we developed an analytical system based on an artificial chewing device coupled with the PTR-ToF-MS. This system allowed a precise dynamic VOC fingerprinting while the food is processed.

1.5 Automated Analysis of Frozen Fruit Tissue

The high biological variability between samples is one of the main risks that has to be considered during the design of an experiment focused on VOC assessment. Therefore, a high number of biological replicates are fundamental for a statistically correct experimental design. The possibility to couple a PTR-ToF-MS instrument with a multipurpose sampler allowed the analysis of more than 200 samples a day with high reproducibility and with reduced lab labor. The limiting factor of this methodology is, however, the restricted volume of the vials that have to be used (20 ml). This restriction can be exceeded by using powdered frozen tissue of the fruit (i.e., cortex or peel).

2 Materials

2.1 Nondestructive Analysis of Intact Fruit

1. Sealed glass jars with a volume of 1000 ml. Each jar lid needs two entries (one inlet and one outlet) with the possibility to be tightly closed.

2. A laboratory water bath.

3. An analytical balance.

4. A zero air producer.

5. PTR-ToF-MS 8000 instrument (Ionicon Analytik GmbH, Innsbruck, Austria).

   The PTR-ToF-MS drift tube is set with the following conditions: 110 °C drift-tube temperature, 2.25 mbar drift pressure, and 550 V drift voltage. This leads to an E/N ratio of about 140 Townsend, where E corresponds to the electric field strength and N to the gas number density.

The sampling time per channel of ToF acquisition was 0.1 ns, amounting to 350,000 channels for a mass spectrum ranging up to m/z = 400. Every single spectrum is the sum of about 28,600 acquisitions lasting 35 µs each, resulting in a time resolution of 1 s.

2.2 Destructive Analysis of Fresh Fruit Tissue

1. Sealed glass jars with a volume of 250 ml. Each jar lid needs two entries (one inlet and one outlet) with the possibility to be tightly closed.

2. A laboratory water bath.

3. A cork borer (diameter of 1.70 cm).

4. Homemade six-blade knife (1 cm distance between blades).

5. A zero air generator.

6. PTR-ToF-MS 8000 instrument (Ionicon Analytik GmbH, Innsbruck, Austria). The PTR-ToF-MS setting is the same described in Subheading 2.1.

2.3 Dynamic Analysis by In Vitro Mastication

1. A laboratory water bath.

2. A cork borer (diameter of 1.70 cm).

3. A zero air generator.

4. PTR-ToF-MS 8000 instrument (Ionicon Analytik GmbH, Innsbruck, Austria). The PTR-ToF-MS setting is the same described in Subheading 2.1.

5. Artificial chewing device [6, 7]. This homemade device is composed of a cylindrical glass cuvette (800 ml) sealed with a cap and a manual notched plunger (Fig. 1). All the device's elements are made of polytetrafluoroethylene. The cuvette cup has two entries: one inlet connected to the zero air generator and one outlet connected to PTR-ToF-MS.

2.4 Automated Analysis of Frozen Fruit Tissue

1. A knife.

2. A basic analytical mill with grinding chamber resistant to liquid nitrogen.

3. Deionized water.

4. Antioxidant solution (sodium chloride, citric acid, and ascorbic acid).

5. Glass vials (volume of 20 ml) equipped with PTFE/silicone septa.

6. A zero air generator.

7. GC autosampler (MPS Multipurpose Sampler, GERSTEL).

8. PTR-ToF-MS 8000 instrument (Ionicon Analytik GmbH, Innsbruck, Austria). The PTR-ToF-MS setting is the same described in Subheading 2.1.

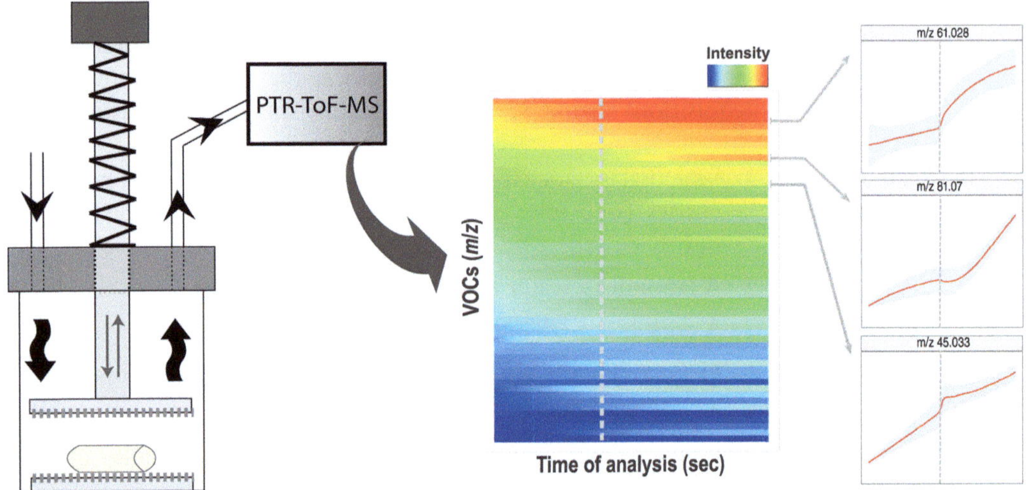

Fig. 1 Schematic representation of the chewing device: cylindrical glass cuvette of 800 ml sealed with a cap and a notched plunger controlled manually. The heat map indicates the dynamic VOC fingerprinting of apple fruit assessed by PTR-ToF-MS coupled with the artificial chewing device. Three examples of VOC release during the dynamic analysis are shown (*m/z* 61.028, 81.07, 45.033). Each graph is divided by a line, at time 0, in two phases, respectively, before and after the chewing moment

3   Methods

3.1 *Nondestructive Analysis of Intact Fruit*

1. A whole intact fruit, without any damages, is weighted and placed into a 1000 ml glass jar and incubated for 30 min at 30 °C into the water bath (*see* **Note 1**).

2. After the incubation, the headspace of the samples is directly connected to the PTR-ToF-MS instrument via a heated PEEK tube (110 °C. 0.055″ diameter) and sampled at a flow rate of 40 standard cm³ per min (sccm). At the same time, the jar is continuously flushed with zero air in order to avoid air contamination and under pressure (*see* **Notes 2–9**).

3. Sampling measurement was performed over 60 cycles resulting in an analysis time of 60 s/sample.

3.2 *Destructive Analysis of Fresh Fruit Tissue*

1. A cylindrical portion of cortex tissue is sampled with the cork borer along the vertical lengthwise plane of the fruit, avoiding the core portion with seeds.

2. From this cylinder, five identical disks with a diameter of 1.70 cm and 1 cm thick are cut with a homemade six-blade knife. Adopting this strategy, we avoid any possible influence of fruit size, and we have tissue samples of different fruit regions.

3. These five disks are placed into a 250 ml glass jar and incubated for 30 min at 30 °C into the water bath (*see* **Note 1**).

4. After the incubation, the headspace of the samples is directly connected to the PTR-ToF-MS instrument via a heated PEEK tube (110 °C. 0.055″ diameter) and sampled at a flow rate of 40 sccm. At the same time, the jar is continuously flushed with zero air in order to avoid air contamination and under pressure (*see* **Notes 2–9**).

5. Sampling measurement was performed in 60 cycles resulting in an analysis time of 60 s/sample.

3.3 Dynamic Analysis by In Vitro Mastication

1. A cylindrical portion of cortex tissue, with a diameter of 1.70 cm and a length of 5 cm, is sampled with the cork borer along the vertical lengthwise plane of the fruit, avoiding the core portion with seeds.

2. This cortex tissue is placed inside the artificial chewing device, constantly maintained at 36 °C into the water bath.

3. Before crushing the headspace VOC concentration of the apple flesh cylinder was measured for 60 s. The chewing was performed by pressing the notched plunger five times within 10 s. VOC analysis continued for 120 s following mastication.

4. The headspace of the samples is directly connected to the PTR-ToF-MS instrument via a heated PEEK tube (110 °C. 0.055″ diameter) and sampled at a flow rate of 40 sccm. At the same time, the jar is continuously flushed with zero air in order to avoid air contamination and under pressure inside the jar (*see* **Notes 2–9**).

3.4 Automated Analysis of Frozen Fruit Tissue

1. The antioxidant solution is prepared mixing 100 ml of deionized water with 40 g of sodium chloride, 0.5 g of ascorbic acid, and 0.5 g of citric acid.

2. Fruit tissue (i.e., cortex or peel) is rapidly cut in small pieces with a sharp knife and immediately frozen into liquid nitrogen. This material can be stored at −80 °C for several weeks before the PTR-ToF-MS analysis.

3. Before being analyzed, this frozen material has to be ground using an analytical mill with the grinding chamber maintained at low temperature by using liquid nitrogen.

4. One gram of powdered frozen sample is immediately inserted into 20 ml glass vials also maintained at low temperature by liquid nitrogen (*see* **Note 1**).

5. One milliliter of the antioxidant solution is added to this sample.

6. These prepared samples can be preserved at 4 °C till the analysis (maximum 24 h).

7. Sampling measurement was performed over 60 cycles resulting in an analysis time of 60 s/sample. Each measurement was con-

ducted automatically after 20 min of sample incubation at 40 °C by using an adapted GC autosampler (MPS Multipurpose Sampler, GERSTEL), and it lasted for about 2 min. During measurements 100 sccm of zero air was continuously injected into the vial, through a needle; the outflow was instead delivered via Teflon fittings to the PTR-ToF-MS through a second heated needle (40 °C) (*see* **Notes 2–9**).

3.5 *Data Analysis*

The analysis of PTR-ToF-MS spectral data (Fig. 2) proceeded correcting offline the count loss (due to the ion detection dead time) through Poisson statistics, following the method reported by Cappellin et al. [8], while the internal calibration was performed according to the procedure described in Cappellin et al. [9] (*see* **Notes 10–20**). Such approach allowed a mass accuracy higher than 0.001 Th to be achieved sufficient for the sum formula determination in our case. Compound annotation was carried out comparing the spectral profile with fragmentation data of reference standards. Noise reduction, baseline removal, and peak intensity extraction were performed according to Cappellin et al. [9], using modified Gaussians to fit the peak shapes. Absolute headspace VOC concentrations, expressed in ppbv (parts per billion by volume), were calculated from peak intensities according to the formula described by Lindinger et al. [2]. A constant reaction rate coefficient of 2×10^{-9} cm^3/s was used in the calculations, introducing a systematic error of up to 30% that can be accounted for if the actual rate coefficient is known [10].

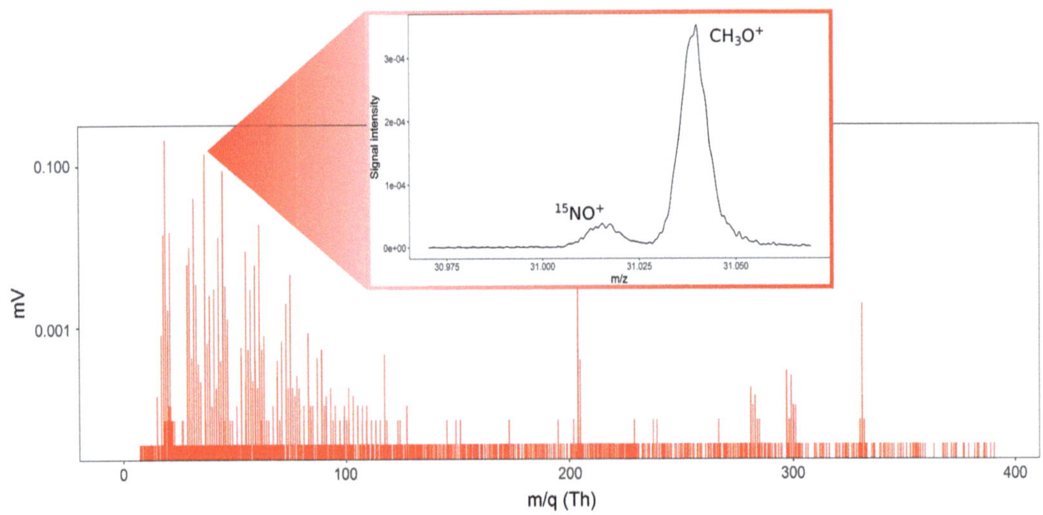

Fig. 2 Example of PTR-ToF-MS spectrum of apple headspace. The main figure represents a sample spectrum in the mass region between 1 and 420 Th, while the second figure enlarges the region around a selected nominal mass peak (*m/z* 31)

4   Notes

To measure with PTR-ToF-MS:

1. Put glass vials and caps in oven at 50 °C for at least 2 days to remove volatiles from a silicon/PTFE septum.

2. Check if you have enough source of primary ion for long experiments.

3. Check if the flow inside the vial would not raise a fine powder or liquid from the surface of the sample.

4. Be sure that each sample VOC would not provoke primary ion depletion. If yes, reduce the quantity, dilute the sample with liquid (i.e., water), or mix the sample headspace with an inert gas (Ar or N_2).

5. Wait enough time between one measurement and another in case of sticky compounds (i.e., α-farnesene for apple fruit). You can even raise the flow in order to clean the PEEK tube with clean air. If you measure with autosampler, you should make some trials before.

6. Control the drift pressure and avoid under- or overpressure in the sample that you have to measure. The wrong drift pressure can cause the problems with VOC protonation.

7. Do not leave the instrument with the ion source switched on if you do not measure anything. It will dirty ion source.

8. Do not smoke or use perfume when you prepare samples or measure the lab air.

9. If it is possible, follow the experiment remotely when you are measuring with autosampler.

To analyze PTR-ToF-MS data:

10. HDF5 file (.h5) is the standard of data acquisition from PTR-ToF-MS.

11. Prepare a log file which consists of "Sample_name," "File_name," "First_spectrum," and "Last_spectrum." The beginning and the end of an experiment can be recorded manually during online measurements. However, if some automation was applied (autosampler, multiple valve control, etc.), it is possible to retrieve these intervals also from HDF5 files in an automatic way.

12. Determine a list of calibration peaks. To become a calibration peak, a mass peak should be single in the window $m/z \pm 0.5$ and have a good intensity (10^2–10^3 counts). We also need to know the theoretical m/z of this mass peak. It can be produced by the instrument (H_3O^+, NO^+, etc.) or derived from the matrix (acetone, acetaldehyde, etc.).

13. Determine a list of test peaks which can be equal to calibration peaks or can be enlarged by other peaks that you want to test.

14. Calibrate files according to the log file and calibration and test peaks.

15. Check the calibration errors which are calculated as the absolute maximum difference of each measurement between the theoretical m/z and the m/z obtained during calibration.

16. If the calibration errors are under the threshold (0.001), it is possible to proceed with the extraction of mass peak concentrations. If not, the calibration peaks should be changed, reduced, or added with new ones, and the calibration should be rerun.

17. For this extraction, it is necessary to have a log file, peaks used for optimization of a general peak shape and another list of peaks for calculation of resolution. These peaks are usually chosen from the calibration list. Optionally it can be downloaded also the peak structure which contains the list of all mass peaks which were extracted for a similar experiment.

18. There are two options of concentration calculation such as online, where concentration for each mass peak is recorded for each spectrum, and average (so-called headspace), where the average concentration of each mass peak for the interval determined in a log file is calculated.

19. It is very useful to optimize and select the mass peaks which will be extracted.

20. It is possible to extract the counts per second for each mass peak and also the concentration in ppbv. For the latter one, it is important to specify the primary ion which was used for the experiment.

References

1. Biasioli F, Yeretzian C, Märk TD et al (2011) Direct-injection mass spectrometry adds the time dimension to (B)VOC analysis. TrAC Trends Anal Chem 30:1003–1017

2. Lindinger W, Hansel A, Jordan A (1998) On-line monitoring of volatile organic compounds at pptv levels by means of proton-transfer-reaction mass spectrometry (PTR-MS) medical applications, food control and environmental research. Int J Mass Spectrom 173:191–241

3. Biasioli F, Gasperi F, Yeretzian C, Märk TD (2011) PTR-MS monitoring of VOCs and BVOCs in food science and technology. TrAC Trends Anal Chem 30:968–977

4. Yeretzian C, Jordan A, Lindinger W (2003) Analysing the headspace of coffee by proton-transfer-reaction mass-spectrometry. Int J Mass Spectrom 223–224:115–139

5. Jordan A, Haidacher S, Hanel G et al (2009) An online ultra-high sensitivity Proton-transfer-reaction mass-spectrometer combined with switchable reagent ion capability (PTR + SRI − MS). Int J Mass Spectrom 286:32–38

6. Farneti B, Khomenko I, Cappellin L et al (2015) Dynamic volatile organic compound fingerprinting of apple fruit during processing. Lebenson Wiss Technol 63:21–28

7. Farneti B, Algarra Alarcón A, Cristescu SM et al (2013) Aroma volatile release kinetics of tomato genotypes measured by PTR-MS following artificial chewing. Food Res Int 54:1579–1588

8. Cappellin L, Biasioli F, Granitto PM et al (2011) On data analysis in PTR-TOF-MS: from raw spectra to data mining. Sensors Actuators B 155(1):183–190

9. Cappellin L, Biasioli F, Schuhfried E et al (2011) Extending the dynamic range of proton transfer reaction time-of-flight mass spectrometers by a novel dead time correction. Rapid Commun Mass Spectrom 25(1):179–183

10. Cappellin L, Soukoulis C, Aprea E et al (2012) PTR-ToF-MS and data mining methods: a new tool for fruit metabolomics. Metabolomics 8(5):761–770

Chapter 16

LC–MS Untargeted Protocol for the Analysis of Wine

Panagiotis Arapitsas and Fulvio Mattivi

Abstract

This chapter describes a protocol for the analysis of the metabolomic fingerprint of wine by liquid chromatography-mass spectrometry. The straightforward, optimized sample preparation procedure is limited to a single-step dilution with water or acetonitrile. The separation of wine analytes is carried out by two columns with orthogonal selectivity, including both reversed-phase (C18) and hydrophilic interaction (HILIC) chromatography, while the detection is assured by a high-resolution quadrupole time-of-flight mass spectrometer operating in negative and positive electrospray ionization mode, in order to obtain four different chromatograms for each sample. This validated protocol, or parts of it, could be applied in several oenological topic experimental designs, including wine quality and wine authenticity.

Key words *Vitis vinifera*, Grape, Food, Holistic, Metabolomics, Mass spectrometry, Liquid chromatography, HILIC

1 Introduction

The ultimate scope of analytical chemistry is to obtain the quantitative characterization of the chemical composition and the elucidation of the structure of all chemical entities present in a given matrix. This is further complicated considering that in living beings, the chemical composition is a dynamic concept, modulated in time.

For many decades, targeted methods have been widely used and taught in university courses of analytical chemistry. Their application is very advanced, and the knowledge and expertise for method development-validation are well established. However, these methods are limited to a small number of known, predefined analytes. Metabolomics developed and evolved as a consequence of the need to obtain comprehensive characterization of the organic molecules in any biological system [1]. In contrast to the targeted methods, in metabolomics, the aim is to achieve the widest possible metabolic coverage in an unsupervised manner, including the compounds so far unknown. In the targeted approach, the majority of

Georgios A. Theodoridis et al. (eds.), *Metabolic Profiling: Methods and Protocols*, Methods in Molecular Biology, vol. 1738, https://doi.org/10.1007/978-1-4939-7643-0_16, © Springer Science+Business Media, LLC, part of Springer Nature 2018

the metabolites present in the matrix are ignored or discarded during the cleanup, the method being entirely focused on the best quantitation of the target compounds, while a basic assumption in the untargeted approach is the desire to monitor as many compounds as possible (ideally all). In consequence, the measured metabolites are by definition not predefined, and the method development and validation follow a different workflow in respect to the targeted analysis [2–5].

Due to the current instrumental limitations and the large chemical diversity, it is not possible to register all the metabolome of one sample with a single analysis, so often complementary platforms are used. The LC-MS techniques are more widely used because they can offer a broader coverage, by combining reversed (semipolar metabolites) with normal-phase (polar metabolites) chromatography and positive with negative electrospray ionization mode in MS. Such an approach could be also further complemented with a GC-MS protocol to cover the volatiles and an adequate LC-MS method to cover the apolar (lipids) metabolites [3, 4, 6, 7].

For the development of a metabolomic fingerprint method, the critical issue to manage is to register the maximum number of features (chromatographic peaks or pairs of m/z with retention time) by maintaining the same sensitivity, retention time, and peak shape during the analysis of all the samples of a sequence. This requires substantial changes in how an analytical chemistry protocol is built, since the usual process of calibration against one or more chemical standard is usually error prone or ineffective in untargeted metabolomics. This is due (1) to the huge chemical diversity of the analytes and (2) to the heavy load of contaminants entering the instruments when the aim of the analysis is the whole metabolome.

Delicate topics to be considered are also the analysis in the same sequence of all the samples of a specific experiment (if possible), the biological variability, and the use of the quality control samples. Important parameters for the targeted methods, such as limits of detection and quantification, matrix effect, and linear range, are difficult—if not impossible—to validate for the development of an untargeted method [5]. We do not want to suggest that they are impossible to achieve within an untargeted method, rather that more robust quantitative data can be obtained developing a targeted quantitative method for a selected number of metabolites, usually defined as putative biomarkers, previously identified via untargeted experiment. The aim of this chapter is to describe how to setup an untargeted analytical protocol which can be used to explore the wine metabolome, producing a validated dataset suitable for further data analysis, in search of the putative biomarkers.

This protocol describes the various steps of the sample preparation, sample analysis, and data analysis in order to obtain four complementary chromatographic traces from wine samples:

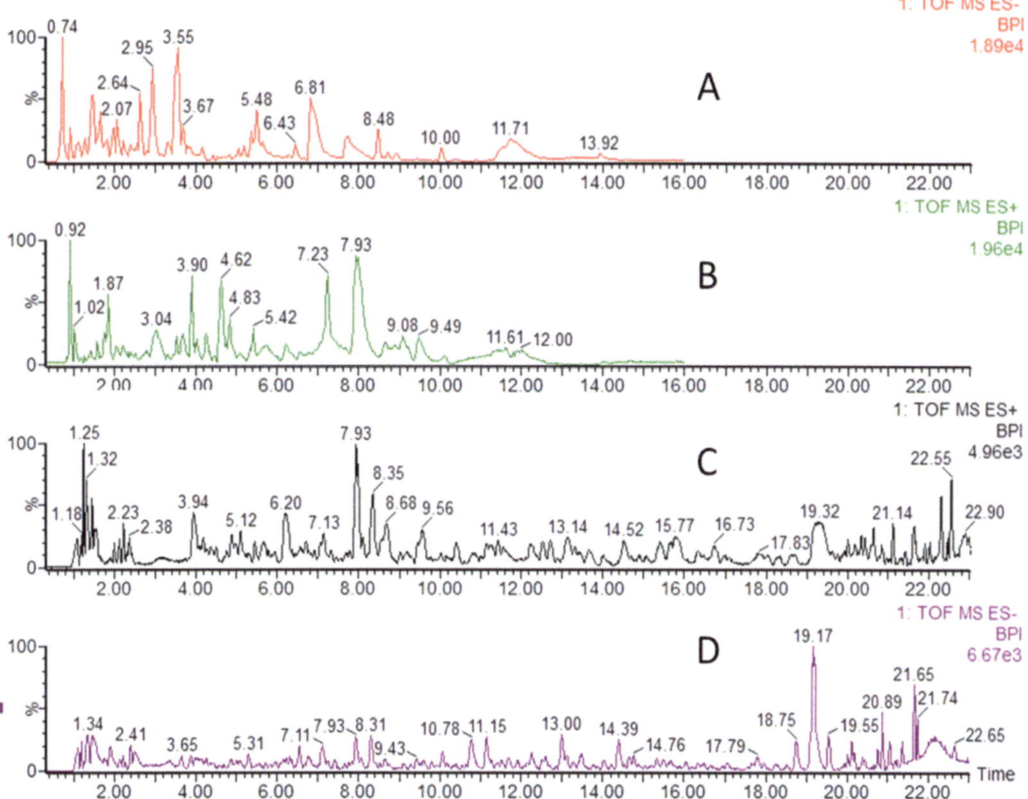

Fig. 1 Typical BPI LC-MS chromatogram of the same wine sample analyzed in four different modes: (**a**) HILIC UPLC-QTOF MS ESI–, (**b**) HILIC UPLC-QTOF MS ESI+, (**c**) reversed-phase UPLC-QTOF MS ESI+, and (**d**) reversed-phase UPLC-QTOF MS ESI–. From these chromatograms, it is clear that each mode covers different parts of the wine metabolome, although some metabolites would be registered from more than one platform

reversed-phase (RP) and hydrophilic interaction (HILIC) ultrahigh-performance liquid chromatography (UPLC) coupled to a high-resolution quadrupole time-of-flight mass spectrometer (QTOF MS) in negative (ESI–) and positive (ESI+) electrospray ionization modes (Fig. 1). Since for metabolomics analysis it is crucial to analyze all the samples within the same sequence, under a randomized order and with the use of quality control (QC) samples, this protocol is applicable to a group of samples and not just one sample. The number of samples, i.e., the size of the experiment, is usually a compromise between the desire to increase, in the "fat" data matrix, the number of samples which is always lower than the number of variables measured and at the same time to prevent the column contamination due to injection of multiple samples within the sequence.

This protocol was already validated and used in wine-related projects [8–11], and here we suggest it could be applied in experimental designs dealing almost with any oenological

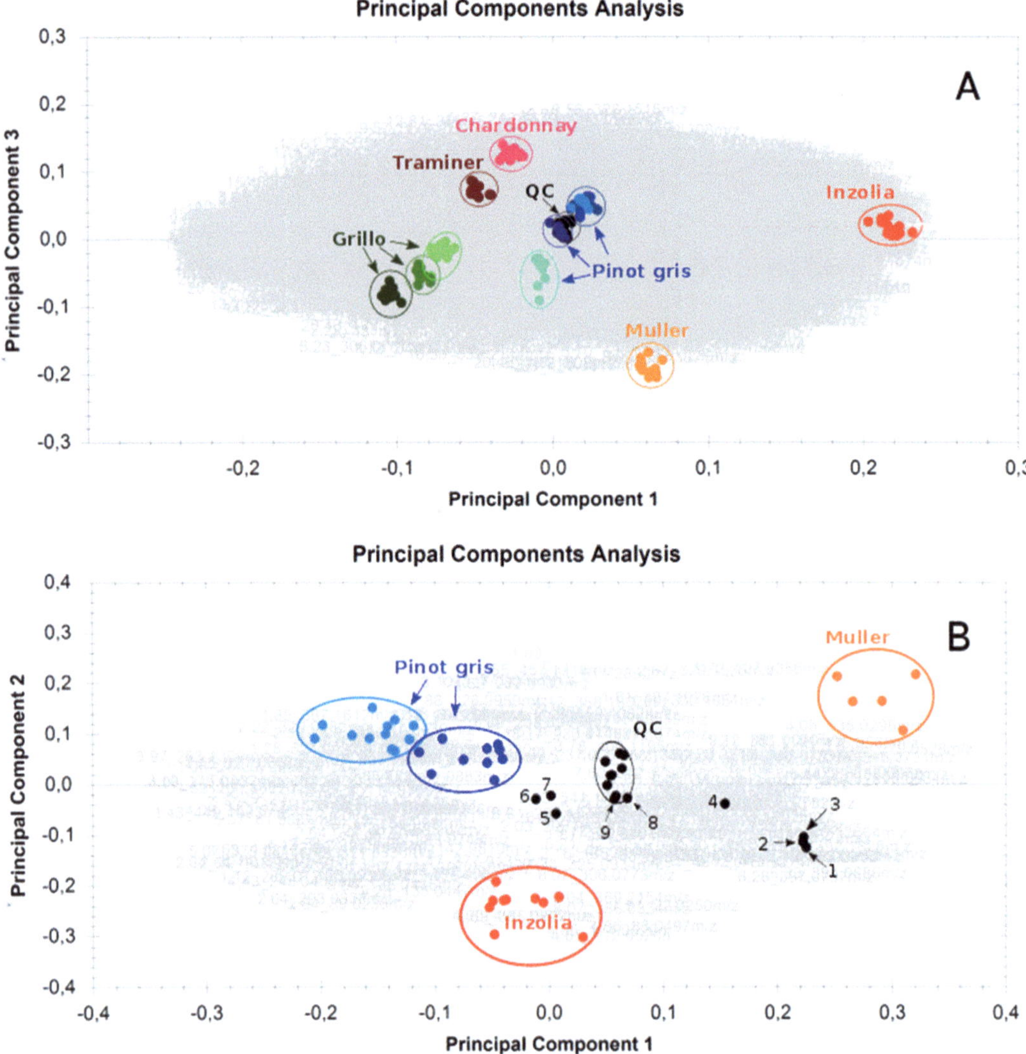

Fig. 2 (**a**) PCA plot of a reversed-phase LS-MS analysis, produced with the described protocol. The experimental design was based in 12 different wines produced by six grape varieties (five Pinot gris, three Grillo, one Chardonnay, one Traminer, one Muller, and one Inzolia wines). For each wine eight different bottles were analyzed. The QC sample injections form a tight cluster in the center of the plot, indicating the quality of the data set. Injections of the same wine biological replicates cluster together, indicating that this method could be used for wine authenticity. (**b**) PCA plot of a HILIC-MS analysis, produced with the described protocol. Also here injections belonging in the same wine clustered together. The numbers which tagged the QCs indicate the order of the injection in the beginning of the sequence. In this experiment, eight injections were necessary to equilibrate the LC-MS; that number is expected for a HILIC-MS untargeted experiment

topic/problem, including wine quality and wine authenticity (Fig. 2). It can be also applied, with small adaptations, to other liquid samples.

2   Material

All solvents (methanol, acetonitrile, isopropanol, chloroform), mobile phase additives (formic acid, ammonia) used should be of LC/MS grade. Water should be ultra-purified (18.2 MΩ), and all the other chemicals used should be of the highest purity (*see* **Note 1**).

2.1   Analytical Instrumentation

The LC system used is a Waters Acquity UPLC, which included a quaternary solvent manager, a sample manager, and a column manager. For the reversed-phase LC, the column used is a Waters Acquity UPLC 1.8 μm 2.1 × 150 mm HSS T3 column and for the normal phase LC/HILIC, a Waters Acquity BEH amide column 2.1 mm × 100 mm × 1.7 μm (HILIC type), equipped with an Acquity UPLC BEH Amide 1.7 μm VanGuard pre-column. The MS system is a Waters Synapt HRMS QTOF MS.

2.2   Solutions

Mobile phase A: a 1 L volumetric flask is half filled with water, then 1 mL of formic acid is added, and finally the flask is filled to volume with water.

Mobile phase B: a 1 L volumetric flask is half filled with methanol, then 1 mL of formic acid is added, and finally the flask is filled to volume with methanol.

A stock solution of $HCOONH_4$ 4 M is prepared in advance and stored at 4 °C. In detail, 12.6 g of $HCOONH_4$ is weighed in a beaker, and about 50 mL of water is added. The solution is stirred for 4 h and then transferred to a 100 mL volumetric flask and filled to volume with water.

For the mobile phase C, 5 mL of the $HCOONH_4$ 4 M stock solution is added to a 1 L volumetric flask containing 950 mL of water, 1 mL of NH_4OH is added, and then the flask is filled to volume (HILIC-LC/MS ESI-). For the mobile phase D, a 1 L volumetric flask is half filled with acetonitrile, then 1 mL of NH_4OH is added, and finally the flask is filled to volume with acetonitrile (HILIC-LC/MS ESI−) (*see* **Notes 2** and **3**).

For the mobile phase E, 5 mL of the $HCOONH_4$ 4 M stock solution is added to a 1 L volumetric flask containing 950 mL of water, 1 mL of formic acid is added, and then the flask is filled to volume (HILIC-LC/MS ESI+). For the mobile phase F, a 1 L volumetric flask is half filled with acetonitrile, then 1 mL of formic acid is added, and finally the flask is filled to volume with acetonitrile (HILIC-LC/MS ESI+)) (*see* **Notes 2** and **3**).

Seal wash solvent is a mixture of 500 mL of water and 500 mL of methanol.

Weak needle wash for the reversed-phase LC (and strong needle wash for the normal phase/HILIC) is a mixture of 100 mL methanol and 900 mL water. Strong needle wash for the reversed-phase LC (and weak needle wash for the normal phase/HILIC) is a mixture of 250 mL methanol, 250 mL acetonitrile, 250 mL iso-propanol, and 250 mL water.

Seal wash solvent is a mixture of 500 mL of water and 500 mL of methanol.

3 Methods

3.1 QTOF MS Parameters

The MS data are collected using different runs in positive and negative ESI mode over a mass range of 50–2000 amu for RP-LC and 30–1000 for HILIC with scan duration of 0.3 s in centroid mode. The instrument operated in W mode. The transfer collision energy and trap collision energy are set at 6 and 4 V. The source parameters are set as follows: capillary 3 kV for positive scan and 2.5 kV for negative scan, sampling cone 25 V, extraction cone 3 V, source temperature 150 °C, desolvation temperature 500 °C, desolvation gas flow 1000 L/h, and nebulizer gas 50 L/h.

External calibration of the instrument is performed at the beginning of each batch of analysis by direct infusion of a sodium formate solution (10% formic acid/0.1 M NaOH/CH_3CN at a ratio of 1/1/8) by controlling the mass accuracy from 50 to 2000 m/z for RP-LC and from 30 to 1000 m/z for HILIC (less than 5 ppm) and mass resolution (over 14,000 FWHM). Lock mass calibration is applied using a solution of leucine enkephalin (0.5 mg/L, m/z 556.2771 for ESI+ and 554.2620 for ESI- mode) at 0.1 mL/min. LC and MS instruments of similar specifications can be used.

3.2 LC Parameters

3.2.1 RP-LC Instrumental Parameters

The column used is held at 40 °C during the analysis, the injection volume is 10 µL, the flow rate is 0.28 mL/min, and the samples are kept at 4 °C throughout the analysis. The multistep linear gradient used is as follows (mobile phases A and B): 0–1 min, 100% mobile phase A isocratic; 1–3 min, 100–90% A; 3–18 min, 90–60% A; 18–21 min, 60–0% A; 21–25.5 min, 0% A isocratic; 25.5–25.6 min, 0–100% A; and 25.6–28 min 100% isocratic (*see* **Note 3**).

3.2.2 HILIC Instrumental Parameters

The column and pre-column are held at 50 °C during the analysis, the injection volume is 10 µL, the flow rate is 500 µL/min, and the samples are kept at 4 °C throughout the analysis. For the ESI- MS, the mobile phases C and D are used, and the multistep gradient elution is as follows: 0–1 min, 88% mobile phase C; 1–2 min, 88–80% of C; 2–10 min, 80–72% C; 10–11 min, 72–50% C; 11–13 min, 50% C; 13–13.1 min, 50–88% C; and 13.1–18 min, 88% C. For the ESI+ MS, the mobile phases E (instead of C) and F (instead of D) are used, under the same multistep gradient (*see* **Note 3**).

3.3 Sample Preparation

Once the sampling is completed and all the samples are in the laboratory, wines should be codified according to a randomized sequence, so the sample preparation and analysis can be completed following this randomized sequence (*see* **Note 4**).

Wines are uncorked under nitrogen atmosphere (*see* **Note 5**), and an aliquot is transferred into a 15 mL amber vial (filled to capacity). Then, again under nitrogen atmosphere, a QC pooled sample is prepared using 0.5 mL of each sample, and this is treated in the same way as the study samples. The QC can be used to optimize the optimum dilution for the LC-MS system used and the number of the samples (*see* **Notes 6** and **7**). For a Waters Acquity UPLC-Synapt QTOF MS system used for this protocol, the following sample preparation is optimized for experimental designs including 60–200 commercial wines (*see* **Notes 8** and **9**).

3.3.1 Sample Preparation for RP-LC/MS

Under N_2 atmosphere, in a 2 mL Eppendorf microtube, 1 mL of each wine is diluted with 1 mL water (*see* **Note 10**), and 20 μL of the internal standard is then added (10 mg *o*-coumaric acid in 10 mL of CH_3OH). Then the solution is filtered with 0.2 μm PTFE filters into a 2 mL amber vial (MS certificated) prior to LC/MS analysis. The same procedure is followed for the blank, but instead of wine 1 mL of water is used.

3.3.2 Sample Preparation for HILIC-MS

Under N_2 atmosphere, in a 2 mL Eppendorf microtube, 0.6 mL of each wine is diluted with 1.2 mL acetonitrile, and 30 μL of the internal standard is then added (14 mg of xanthosine and 25 mg of nicotinic acid in 10 mL of $CH_3OH:H_2O$ (1:1)). Then the solution is filtered with 0.2 μm PTFE filters into a 2 mL amber vial (MS certificated) prior to LC/MS analysis. The same procedure is followed for the blank, but instead of wine 0.6 mL of water is used.

3.4 LC-MS System Start-Up and Pretest

1. Load mobile phases and wash solutions.

2. Clean the ESI source according to the vendor protocol.

3. Prime the mobile phases for 5 min.

4. Connect the column.

5. Flush the column for 10 min at 0.05 mL/min, with 100% B for reversed-phase LC or 100% D/F for normal phase LC.

6. Increase the column oven temperature at 40 °C for RP-LC and 50 °C for HILIC.

7. Set the mobile phase at 50% B—50% A for RP-LC or 80% D/F—20% C/E for HILIC.

8. Calibrate the MS and control the mass accuracy according to the vendor protocol.

9. Bring gradually first the mobile phase % and then the flow rate at the initial LC condition.

10. Load the UPLC method and wait until the back pressure equilibration (psi delta <20).

3.5 Sample Sequence

The samples should be injected according to the randomized order established before the sample preparation (*see* Subheading 3.3). Then the analysis should be made according to the following sequence (*see* **Notes 3, 6–9, 11–15**):

1. First run a blank sample injection.

2. Then run four QC injections for the reversed phase and ten QC injections for the normal phase LC, in order to equilibrate the column (Fig. 2).

3. Now you can start the injection of the real samples, but for every six real sample injections, one QC sample should be injected.

4. The sequence should end with one QC sample and finally one blank sample injection.

3.6 Data Analysis

Data analysis in metabolomics follows a specific but very wide workflow [11, 12] that should include the following steps:

1. Data preprocessing, divided in deconvolution, peak picking, filtering (optional), alignment, and bucketing (binning).

2. Data pretreatment, subdivided into normalization (optional), centering (optional), scaling, managing missing values, and managing outliers.

3. Data processing, employing the use of statistical tools such as multivariate supervised and unsupervised analysis.

4. Data visualization, such as PCA plots (Fig. 2).

5. Data quality validation.

6. Marker detection, annotation, and interpretation.

The researcher can use various informatics tools (one or more) to accomplish the above workflow, based in his/her experience/knowledge and financial support. For example, XCMS online [12] is a valid, free, and widely used tool. In this protocol, we propose the use of the commercial and user-friendly software Progenesis QI (Waters, Nonlinear Dynamics). Other useful open source informatics tools are mzMine [13], MetaDB [11], MetAlign [14], and MetaboAnalyst [15].

Between the objects of this protocol, the marker detection and validation, metabolite annotation, or the hypothesis generation is not included. For this reason, the data analysis will be concluded with the data visualization (Fig. 2) which is important to validate the quality of the data set produced by the instrumental analysis (*see* **Note 15**).

Progenesis QI steps:

1. Make a new experiment, and choose a name for it and a directory to save the files created by Progenesis QI.

2. Set the analysis parameters according to your instrument, the data format (centroided with 17,000 resolution in our case), and ionization polarity of the analysis (positive or negative).

3. Import the raw files of the injections to Progenesis QI (deconvolution and peak picking).

4. Start automatic processing (alignment), by selecting a QC from the middle of the sequence as reference, untick the blank samples (never use blanks or standard mixes for the alignment), and ignore ions eluting before the 1st minute and after the 22nd minute of the chromatography.

5. Inspect visually the alignment but without performing any manual changes.

6. In the experimental design page, divide your samples with respect to the groups of the study (i.e., treatment vs. control or each wine variety in a different group or wines from different zones in different groups).

7. Do the peak picking after the experimental design set up, without changing the parameters. This peak picking is applied to the compounds (the software is grouping isotopic peaks and adducts belonging to one compound) and not to ions as before.

8. After that go directly to the Compound Statistic page to visualize the PCA plot. This is the first quality control of the analysis (data). For a good quality data, the QC samples should group together to a tight cluster (Fig. 2).

4 Notes

1. Please consult all relevant material safety data sheets (MSDS) before use. Some of the chemicals used in this protocol are acutely toxic and carcinogenic. Please use all appropriate safety practices when performing the extraction including the use of engineering controls (fume hood, glovebox) and personal protective equipment (safety glasses, gloves, lab coat, full-length pants, closed-toe shoes).

2. For an efficient hydrophilic interaction chromatography (HILIC) method, analytes often exhibit the strongest retention when they are ionized, thus bases at low to mid-pH and acids at mid- to high pH. This is the reason why we propose, for the HILIC analysis, to combine acidic mobile phases with ESI+ data acquisition and basic mobile phases with ESI− data acquisition.

3. Prepare all the necessary mobile phases 1 day before the analysis (especially for HILIC), and load all the volume in the LC bottles from the beginning. Avoid adding mobile phase during the analysis.

4. It is important to collect, register, and keep track of the details and the characteristics of all samples analyzed. MetaDB [11] is a useful tool for this aim. Register all the meta-data information of the samples according to the minimum reporting standards for plant biology context information in metabolomics studies [16].

5. N_2 atmosphere is important to avoid possible O_2-driven reactions.

6. If the QC sample is a pooled sample, prepared by mixing equal aliquots of each individual sample of the sample set, can provide import help for (a) training, (b) method development/adaptation, (c) column equilibration, (d) data quality control, and (e) marker quality evaluation.

7. Surrogate QC samples can be used in long-term studies which include more than one sequence. Standard mixes are not considered good QC solution in the metabolomics field.

8. For commercial light-bodied white wines, it is possible to perform 200 injections without losing sensitivity. For full-bodied red wines, the maximum injection suggested for this sample preparation/instrumental setup is around 80 injections.

9. In case of a large number of samples (over 200), consider the possibility to divide the sample set in smaller subsets. In this case instrumental and data analysis should be made separately, and then compared for common markers. All sample subsets should have equally divided groups, i.e., same number of control and treatment condition samples, and not one subset with the control samples and one subset with all the treatment samples.

10. Degas the water used for the wine dilution.

11. Clean the LC tubing and pumps with a solution of 25% methanol, 25% acetonitrile, 25% isopropanol, and 25% water every 2 weeks for reversed phase and every 5 days for HILIC.

12. If standard mixes are included in the sequence, inject them (a) in the beginning of the sequence after the blanks and before the QCs and (b) in the end of the sequence after the last QC and before the last blank injection.

13. If blanks or standard mixes are included during the injections of the real samples, run a minimum of two QC injections before continuing with the real sample injections.

14. Monitor during the sequence (a) the LC back pressure, (b) retention time of specific metabolites, (c) mass error for specific metabolites, and (d) the area for specific metabolites.

15. Use data visualization with unsupervised multivariate statistical tools, such as PCA plot, to control the data quality, a step which should be frequently monitored during the sequence of analysis, by examining the QC distribution/clustering in the PCA plot and/or possible outliers in the PCA plot (Fig. 2). If you have to reinject—or re-prepare and reinject—more sample(s), it is better to include it/them in the end of the sequence before the last QC.

References

1. Nicholson JK, Lindon JC (2008) Systems biology: metabonomics. Nature 455:1054–1056

2. Gika HG, Theodoridis GA, Vrhovsek U et al (2012) Quantitative profiling of polar primary metabolites using hydrophilic interaction ultrahigh performance liquid chromatography-tandem mass spectrometry. J Chromatogr A 1259:121–127

3. Theodoridis G, Gika H, Franceschi P et al (2011) LC-MS based global metabolite profiling of grapes: solvent extraction protocol optimisation. Metabolomics 8:175–185

4. Theodoridis GA, Gika HG, Want EJ et al (2012) Liquid chromatography–mass spectrometry based global metabolite profiling: a review. Anal Chim Acta 711:7–16

5. Naz S, Vallejo M, García A et al (2014) Method validation strategies involved in non-targeted metabolomics. J Chromatogr A 1353:99–105

6. Buscher JM, Czernik D, Ewald JC et al (2009) Cross-platform comparison of methods for quantitative metabolomics of primary metabolism. Anal Chem 81:2135–2143

7. Cajka T, Fiehn O (2016) Toward merging untargeted and targeted methods in mass spectrometry-based metabolomics and Lipidomics. Anal Chem 88:524–545

8. Arapitsas P, Speri G, Angeli A et al (2014) The influence of storage on the "chemical age" of red wines. Metabolomics 10:816–832

9. Arapitsas P, Ugliano M, Perenzoni D et al (2016) Wine metabolomics reveals new sulfonated products in bottled white wines, promoted by small amounts of oxygen. J Chromatogr A 1429:155–165

10. Mattivi F, Arapitsas P, Perenzoni D et al (2015) Influence of storage conditions on the composition of red wines - advances in wine research - ACS symposium series. In: ACS symposium series. ACS Publications, Washington, DC, pp 29–49

11. Franceschi P, Mylonas R, Shahaf N et al (2014) MetaDB a data processing workflow in untargeted MS-based metabolomics experiments. Front Bioeng Biotechnol 2:72

12. Smith CA, Want EJ, O' Maille G et al (2006) XCMS: processing mass spectrometry data for metabolite profiling using nonlinear peak alignment, matching, and identification. Anal Chem 78(3):779–778

13. Katajamaa M, Miettinen J, Oresic M (2006) MZmine: toolbox for processing and visualization of mass spectrometry based molecular profile data. Bioinformatics 22:634–636

14. Lommen A (2009) MetAlign: interface-driven, versatile metabolomics tool for hyphenated full-scan mass spectrometry data preprocessing. Anal Chem 81:3079–3086

15. Xia J, Wishart DS et al (2016) Using MetaboAnalyst 3.0 for comprehensive metabolomics data analysis. Curr Protoc Bioinformatics 55:14.10.1–14.10.91. https://doi.org/10.1002/cpbi.11

16. Sumner LW, Amberg A, Barrett D et al (2007) Proposed minimum reporting standards for chemical analysis chemical analysis working group (CAWG) metabolomics standards initiative (MSI). Metabolomics 3:211–221

Part IV

Life Science Applications

Chapter 17

Tissue Multiplatform-Based Metabolomics/Metabonomics for Enhanced Metabolome Coverage

Panagiotis A. Vorkas, M. R. Abellona U, and Jia V. Li

Abstract

The use of tissue as a matrix to elucidate disease pathology or explore intervention comes with several advantages. It allows investigation of the target alteration directly at the focal location and facilitates the detection of molecules that could become elusive after secretion into biofluids. However, tissue metabolomics/metabonomics comes with challenges not encountered in biofluid analyses. Furthermore, tissue heterogeneity does not allow for tissue aliquoting. Here we describe a multiplatform, multi-method workflow which enables metabolic profiling analysis of tissue samples, while it can deliver enhanced metabolome coverage. After applying a dual consecutive extraction (organic followed by aqueous), tissue extracts are analyzed by reversed-phase (RP-) and hydrophilic interaction liquid chromatography (HILIC-) ultra-performance liquid chromatography coupled to mass spectrometry (UPLC-MS) and nuclear magnetic resonance (NMR) spectroscopy. This pipeline incorporates the required quality control features, enhances versatility, allows provisional aliquoting of tissue extracts for future guided analyses, expands the range of metabolites robustly detected, and supports data integration. It has been successfully employed for the analysis of a wide range of tissue types.

Key words Metabolomics, Metabonomics, Metabolic profiling, Metabolic phenotyping, Lipidomics, Tissue, Extraction, Metabolome, Lipidome, Coverage, Multiplatform, NMR, UPLC-MS, MS^E, HILIC

1 Introduction

Metabolomics/metabonomics applications in tissue, as a matrix for studying disease or intervention, come with several advantages, all of which derive from the fact that the target pathology/alterations are directly examined. Firstly, this allows for easier and more relevant mechanistic elucidation, which in turn can provide targets for pharmacotherapeutic intervention. Additionally, differentially produced metabolites can be more easily identified in the corresponding tissue, as they may be metabolized, excreted, diluted or confounded by the complexity of the biofluid after secretion by the tissue. After being identified as differentially produced within the tissue, subsequent follow-up using targeted methods that have

Georgios A. Theodoridis et al. (eds.), *Metabolic Profiling: Methods and Protocols*, Methods in Molecular Biology, vol. 1738, https://doi.org/10.1007/978-1-4939-7643-0_17, © Springer Science+Business Media, LLC, part of Springer Nature 2018

higher sensitivity may be applied in biofluids as a less invasive option. Tissue candidate biomarkers can also be explored in the context of in vivo imaging. Lastly, tissue analysis allows for the mapping of organ-to-organ interactions and crosstalk, which has been proven to be very informative when systemically studying disease [1].

The use of tissue as a matrix for metabolomics/metabonomics applications has increased in the recent years. Papers employing tissue metabonomics have tripled in the past 5 years. Although metabolite extraction methods employ a solvent-based approach that was described several decades back [2, 3], the current trend is diverging from conventional approaches and moving toward more high-throughput settings. Homogenization techniques such as the mortar and pestle are now abandoned, and techniques such as bead beating are favored [4]. At the same time, the traditional bilayer extraction methodology is replaced by consecutive monolayer extractions using solvents with different physicochemical properties. This can provide higher repeatability [5]. Extensive studies in a comprehensive metabolomic setting aiming to identify the optimal solvent systems for organic and aqueous extractions have also been conducted [6].

Here we describe a multiplatform, multi-method workflow which enables metabolic profiling analysis of tissue samples and covers an extended part of the metabolome [7]. This workflow utilizes dual consecutive monolayer extractions. Tissue extracts are subsequently analyzed by untargeted reversed-phase (RP-) and hydrophilic interaction liquid chromatography (HILIC-) ultra-performance liquid chromatography coupled to mass spectrometry (UPLC-MS) and nuclear magnetic resonance (NMR) spectroscopy. RP-UPLC-MS is employed for the analysis of organic extracts (lipid profiling) and HILIC-UPLC-MS and NMR for the profiling of aqueous extracts (polar metabolite profiling). This pipeline incorporates the required quality control features, enhanced versatility, repeatability, and reproducibility. It allows provisional aliquoting of tissue extracts for future analyses (which could be guided by the results of the initial untargeted analyses), supports data integration, and covers an extended part of the metabolome with potential for detecting unknown molecules [8]. It has been successfully employed for the analysis of several tissue types from humans, rodents, and bovines, such as liver, brain, spleen, kidney, breast (normal and cancerous), adipose, arteries and veins, atherosclerotic plaques, tissue segments covering the whole range of the gastrointestinal track, skeletal muscle, heart, and also model organisms such as *C. elegans* [6–12].

---

2   Materials

All solvents used should be LC-MS grade. All solvent mixtures are described as volume parts per solvent. Efforts should be taken to minimize the presence of contaminants in containers used (*see* **Note 1**).

2.1 Tissue Dissection	1. Dry ice. 2. Weighing boats (large for cutting and small for weighing). 3. Balance (≤0.1 mg readability). 4. Single-use scalpels and forceps (*see* **Note 2**).

2.2 Tissue Extraction

1. Bead beating: bead beating tubes (2 mL), filled with 1 mm zirconium beads. Zirconium beads should cover just above the curved bottom of the tube (*see* **Notes 3** and **4**).

2. Bead beater (Bertin Technologies).

3. Solvent mixture for organic extraction: methanol/methyl *tert*-butyl ether (MTBE) (1:3).

4. Solvent mixture for aqueous extraction: water/methanol (MeOH) (1:1)

5. Centrifuge (e.g., Eppendorf).

6. Eppendorf tubes 1–2 mL (*see* **Note 5**).

7. Solvent mixture evaporation using vacuum concentrator working at 30 °C (2 h for organic and 2.5 h for aqueous extraction) (*see* **Note 6**).

2.3 Analysis

2.3.1 RP-Lipid Profiling

1. Reconstitution solvent mixture: water/acetonitrile (ACN)/isopropanol (IPA) (1:1:3).

2. Multi-sample vortex mixer.

3. Centrifuge (Eppendorf).

4. Vials: deactivated clear glass 12 × 32 mm screw neck total recovery vial, large glass 20 mL vials.

5. Mobile phase A: ACN/water (60:40), 10 mM ammonium formate, and 0.1% formic acid. For 2 L of mobile phase A, add 800 mL of water followed by the addition of 1.26 g of ammonium formate. Swirl and sonicate until the salt has been fully dissolved. Proceed with the addition of 2 mL of formic acid. Swirl briefly. Add 1200 mL of acetonitrile and degas for 5 min by sonication.

6. Mobile phase B: IPA/ACN (90:10) and 0.1% formic acid. For 2 L of mobile phase B, add 1800 mL of isopropanol, 200 mL of acetonitrile, and 2 mL of formic acid. Degas by brief sonication.

7. UPLC separation is performed using an Acquity UPLC System (Waters Corp., USA).

8. Column: Acquity UPLC CSH C18 2.1 × 100 mm, 1.7 µm, column (Waters Corp, USA).

9. Weak wash solvent mixture: ACN/water (60:40).

10. Strong wash solvent mixture: IPA/ACN (90:10).

11. Seal wash: add 125 mL of IPA to 2.5 L water.

12. MS: Xevo G2 QTof (Waters MS Technologies, UK) with an electrospray ionization (ESI) source.

2.3.2 HILIC-MS of Aqueous Extract

1. Reconstitution solvent mixture: acetonitrile (ACN)/water (95:5).

2. Multi-sample vortex mixer.

3. Centrifuge (e.g., Eppendorf).

4. Sonicator.

5. Vials: LCGC certified clear glass 12 × 32 mm screw neck vial, 150 µL inserts (deactivated).

6. Mobile phase A: acetonitrile (ACN)/water (95:5), 10 mM ammonium acetate, and 0.1% formic acid. For 2 L of mobile phase A, add 100mL of water, followed by 1.54 g of ammonium acetate. Sonicate until the salt has been fully dissolved. Add 2 mL of formic acid and briefly swirl the bottle. Proceed by gradually adding 100 mL of acetonitrile and sonicating for 10 min. Then, add another 200 mL of acetonitrile and 5 min sonication. Continue with 400 mL acetonitrile/5 min sonication, and lastly add 1200 mL of acetonitrile and sonicate for a final 5 min.

7. Mobile phase B: ACN/water (50:50), 10 mM ammonium acetate, and 0.1% formic acid. For 2 L of mobile phase B, in 1 L of water, add 1.54 g of ammonium acetate. Sonicate until the salt has been fully dissolved. Add 2 mL of formic acid and swirl briefly. Proceed by gradually adding 1 L of acetonitrile and sonicate for 5 min.

8. UPLC separation is performed using an Acquity UPLC System (Waters Corp., USA).

9. Column: Acquity UPLC BEH HILIC 2.1 × 100 mm, 1.8 µm, column (Waters Corp, USA).

10. Weak wash solvent mixture: ACN/water (95:5).

11. Strong wash solvent mixture: ACN/water (50:50).

12. Seal wash: add 125 mL of IPA to 2.5 L water.

13. MS: Xevo G2 QTof (Waters MS Technologies, UK) with an electrospray ionization (ESI) source.

2.3.3 NMR Spectroscopy

1. Reconstitution buffer: for 1 L of buffer, mix 500 mL of high purity water with 21.73 g Na_2HPO_4 (anhydrous), 2.63 g NaH_2PO_4 (anhydrous), and 0.1 g sodium 3-trimethylsilyl-1-[2,2,3,3,-d_4] propionate (TSP) (0.01% m/v). Mix well and then add 200 mL of D_2O. Mix until salts are completely dissolved. Sonicate the solution or use magnetic spinner if necessary. Adjust to pH 7.4 with NaOH crystals or HCl (high

concentration). Transfer the mix to a 1 L volumetric flask and fill up to 1 L with high purity water. Readjust the pH to 7.4. Consider the addition of a bacteriostatic agent (*see* **Note 7**).

2. Tubes: NMR tubes with an outer diameter of 5 mm.

3. NMR instrument: a 600 MHz NMR spectrometer (Bruker BioSpin).

2.4 Data Deconvolution and Processing

2.4.1 UPLC-MS

Centroiding is performed using MassLynx (Waters). The DataBridge software (MassLynx, Waters) is used for conversion of files from MassLynx (UPLC-MS analysis) to NetCDF format. The XCMS package (version 1.46.0) using R (version 3.2.2) is employed for data deconvolution.

2.4.2 NMR Spectroscopy

MATLAB (version 2014a) programming language is utilized for importing and processing of NMR data, using in-house developed scripts.

2.5 Statistical Analysis

For multivariate statistical analysis, such as principle component analysis (PCA) and orthogonal projection to latent structures (OPLS), SIMCA version 14.0.0 is used. For *t*-test and fold change calculations, MS Excel is used. Spearman correlations are calculated using R version 3.2.2. Correlation networks are constructed using Cytoscape version 3.4.0 (Cytoscape Consortium) [13].

2.6 Pathway Mapping

For pathway mapping the KEGG database [14] and Ingenuity Pathway Analysis software (QIAGEN) is used.

---

3 Methods

An overview of the procedure is summarized in Fig. 1.

According to the origins of the tissue used, severe biohazards might exist. Take every required precaution.

3.1 Randomization

The MS Excel random number generating function, RAND, is used for sample order randomization. After generating the random number using the function, sample labels were sorted according to the generated random number.

3.2 Tissue Handling and Dissection

1. Tissues samples must be stored at −80 °C.

2. Maintain tissue samples in dry ice when outside of the freezer (during transporting, handling etc).

3. Use single-use, disposable scalpels and forceps (*see* **Notes 2** and **8**).

4. Cut each tissue on a new/clean large container (large weighing boat). It is recommended to place the container with the tissue

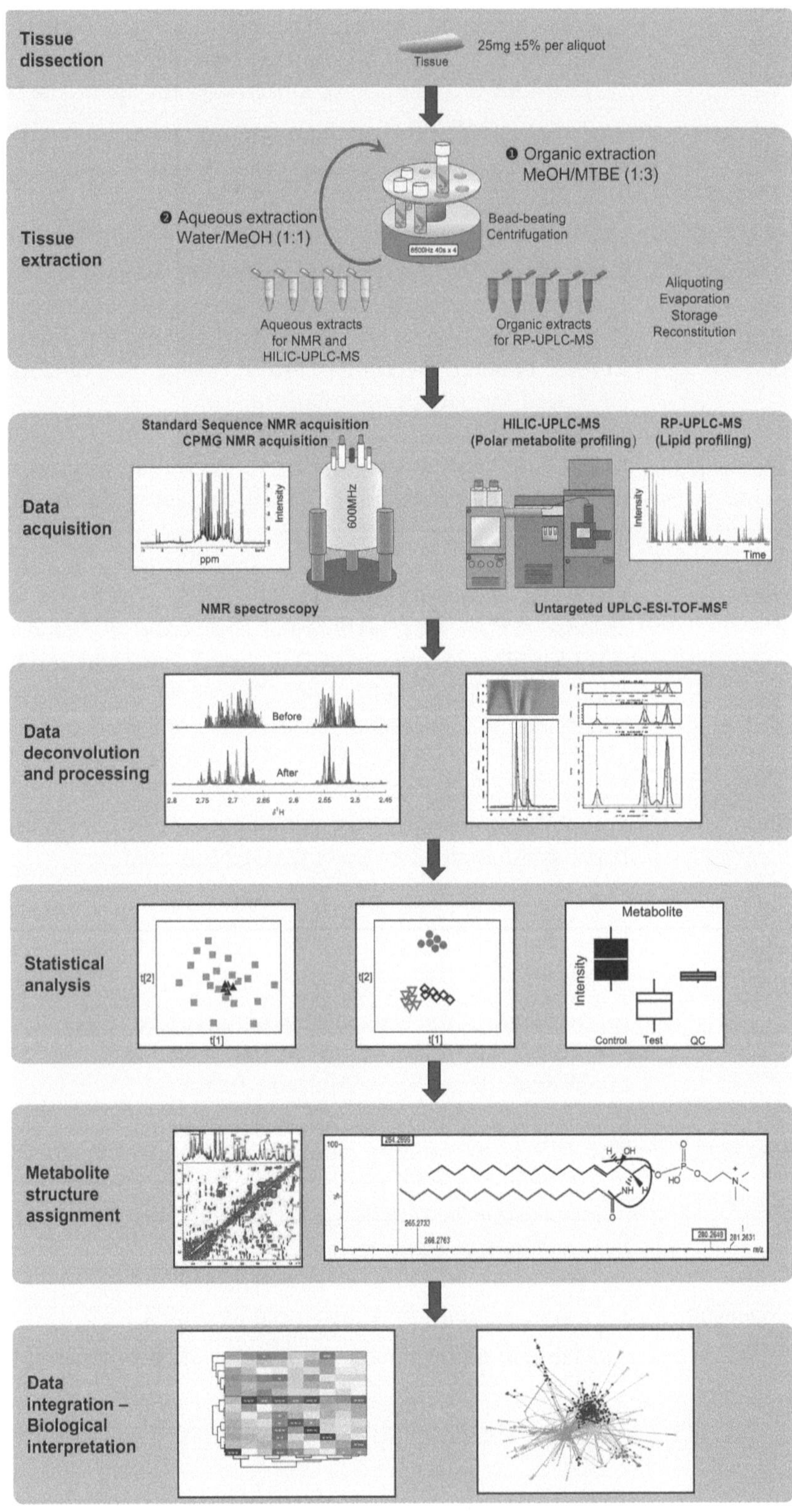
Tissue
dissection
25mg ±5% per aliquot
Tissue
Tissue
extraction
❶ Organic extraction
MeOH/MTBE (1:3)
❷ Aqueous extraction
Water/MeOH (1:1)
Bead-beating
Centrifugation
6500Hz 40s x 4
Aliquoting
Evaporation
Storage
Reconstitution
Aqueous extracts
for NMR and
HILIC-UPLC-MS
Organic extracts
for RP-UPLC-MS
Data
acquisition
Standard Sequence NMR acquisition
CPMG NMR acquisition
HILIC-UPLC-MS
(Polar metabolite profiling)
RP-UPLC-MS
(Lipid profiling)
Intensity
600MHz
ppm
Intensity
Time
NMR spectroscopy
Untargeted UPLC-ESI-TOF-MS
Data
deconvolution
and processing
Before
After
2.8 2.75 2.7 2.65 2.6 2.55 2.5 2.45
δ¹H
Statistical
analysis
t[2]
t[1]
t[2]
t[1]
Metabolite
Intensity
Control Test QC
Metabolite
structure
assignment
100
264.2690
265.2732
266.2763
286.2640 281.2631
m/z
Data
integration –
Biological
interpretation

on top of dry ice, in order to maintain the tissue sample at a low temperature. Transfer and weigh the tissue sample in a new clean weighing boat (use the empty weighting boat to tare the balance, prior to loading the tissue sample). One aliquot should correspond to 20–25 mg of wet tissue mass. One aliquot should be considered the optimum material for most MS applications and two aliquots for NMR spectroscopy (although as little as half the recommended mass can be used). The total amount of the required wet tissue should correspond to (be multiplied by) the number of aliquots required for the methods/platforms employed (20–25 mg x required aliquots). Samples should not differ by more than 5% of their total weight. In order to follow the protocol described here, 60–75 mg ± 5% of wet tissue is required (3 aliquots). Other options are available for difficult to cut tissue or when cutting should be avoided altogether (*see* **Note 9**).

5. Transfer the dissected/weighed tissue into the bead beating tube (preloaded with beads).

6. Include blank samples to test for contamination induced by the extraction procedure (extraction blanks). Set up bead beating tubes without tissue sample. The number of recommended extraction blanks used should be ≥7 and ≥5% of the total number of samples.

3.3 *Tissue Extraction*

3.3.1 Organic Extraction (Round 1)

1. Randomize samples.

2. Add 1500 μL of the organic extraction solvent to the bead beating tubes loaded with the weighed tissue and/or beads (samples/blanks) (*see* **Notes 9–12**).

3. Load the bead beater and initiate the vibration sequence at 6500 Hz for 40 s. Repeat for two to four cycles. Each cycle should be separated by placing the samples in dry ice for 5 min (*see* **Note 13**).

4. Centrifuge at 20,000 rcf for 30 min at 4 °C.

5. Decant the number of aliquots corresponding to the amount of tissue used into Eppendorf tubes. For 75 mg of wet tissue, the organic extraction solvent should be divided into three aliquots of 400 μL.

6. Produce extraction pooled samples by combining a small amount of the remaining tissue extracts by decanting an equal volume from each bead beating tube in a large container.

Fig. 1 An overview of the tissue metabolomics/metabonomics multiplatform procedure. The procedure is comprised of seven basic steps: (1) tissue dissection, (2) metabolite extraction from the tissue sample (two consecutive extractions: organic, round 1, followed by aqueous, round 2), (3) data acquisition (NMR, HILIC-, and RP-UPLC-MS), (4) data deconvolution and processing, (5) statistical analysis, (6) metabolite structure assignment, and (7) data integration and biological interpretation. MeOH, Methanol; MTBE, methyl *tert*-butyl ether

Proceed by aliquoting the same volume as with individual samples (**step 5**) into Eppendorf tubes. If 75 mg of tissue is used, then the pooled extracts should be aliquoted in 400 μL (*see* **Note 14**).

7. Proceed to complete solvent evaporation of the aliquots using a vacuum concentrator working at 30 °C, for 2 h (*see* **Notes 15–17**).

8. Ensure the bulk of the organic extraction solvent has been completely removed from the bead beating tube before performing the aqueous extraction.

3.3.2 Aqueous Extraction (Round 2)

1. Randomize samples (optional; *see* **Note 18**).

2. Add 1500 μL of the aqueous extraction solvent to the bead beating tubes loaded with weighed tissue and/or beads (samples/blanks) (*see* **Note 10–12**).

3. Load in bead beater vibrating at 6500 Hz for 40 s, for two to four cycles. Cycles are separated by freezing of the samples in dry ice until the extraction solvent mixture freezes (~5 min) (*see* **Note 19**).

4. Centrifuge at 20,000 rcf for 30 min at 4 °C.

5. Decant the number of aliquots corresponding to the amount of tissue used into Eppendorf tubes. For 75 mg of tissue, the aqueous extraction solvent should be divided in three aliquots of 400 μL. Two aliquots should be considered the optimal amount for NMR analysis of aqueous extracts (*see* **Notes 12 and 20**).

6. Produce extraction pooled samples by combining a small amount of the remaining solvent (of the homogenized tissue) by decanting an equal volume from each bead beating tube in a large container. Proceed by aliquoting the same volume as with individual samples into Eppendorf tubes (as with **step 5**). For 75 mg of tissue, 400 μL should be aliquoted (*see* **Note 14**).

7. Proceed to complete solvent evaporation of the aliquots using a vacuum concentrator working at 30 °C, for 2.5 h (*see* **Notes 15 and 16**) .

3.4 UPLC-MS Analysis and Data Treatment

3.4.1 Reconstitution

Lipid Profiling

1. Add 200 μL of the lipid profiling reconstitution solvent mixture to the Eppendorf tubes of the extracts and blanks (*see* **Note 21**).

2. Vortex vigorously for 1 min.

3. Centrifuge for 20 min at 20,000 rcf at 4 °C.

4. Decant the supernatant into an LC-MS total recovery vial.

5. Combine 50 μL of the reconstitute to produce the QC pooled sample (*see* **Notes 14 and 22**).

6. Use the QC or extraction pooled samples to perform a dilution series by diluting using the reconstitution solvent mixture (*see* **Note 23**) .

HILIC

1. Add 120 μL of the HILIC reconstitution solvent mixture in the Eppendorf tubes of the extracts and blanks (*see* **Note 21**).

2. Vortex vigorously for 1 min.

3. Sonicate for 5 min.

4. Vortex vigorously for 1 min.

5. Centrifuge for 20 min at 20,000 rcf at 4 °C.

6. Combine approximately 20 μL of the reconstitute to produce the QC pooled sample (*see* **Notes 14** and **22**).

7. Decant the remainder of the supernatant into an LCGC vial loaded with deactivated inserts.

8. Use the QC or extraction pooled samples to perform a dilution series by diluting with the reconstitution solvent mixture (*see* **Note 23**) .

3.4.2 UPLC-MS

UPLC Conditions

RP-lipid profiling

The suggested binary solvent manager parameters are weak wash volume, 1000 μL; strong wash volume, 1000 μL; seal wash frequency, 1 min; column temperature, 55.0 °C; autosampler temperature, 8.0 °C; partial loop with needle overfill option; needle overfill flush, 4 μL; and injection volume 3 μL for positive polarity and 10 μL for negative polarity mode (*see* **Note 21**).

The elution gradient is set as follows: 60–57% A(0.0–2.0 min), 57–50% A (2.0–2.1 min; curve 1), 50–46% A(2.1–12.0 min), 46–30% A (12.0–12.1 min; curve 1), 30–1%A (12.1–18 min), 1–60% A (18.0–18.1 min), and 60% A (18.1–20.0 min). Flow rate is maintained at 0.4 mL/min.

Characteristic chromatograms with detection in positive and negative polarity mode are illustrated in Fig. 2.

HILIC

The suggested binary solvent manager parameters are weak wash volume, 600 μL; strong wash volume, 200 μL; seal wash frequency, 1 min; column temperature, 40.0 °C; autosampler temperature, 4.0 °C; partial loop with needle overfill option; needle overfill flush, 4 μL; and injection volume 5 μL for both polarity modes (*see* **Note 21**).

The elution gradient and flow rate are set as follows:

99% A (0.0–2.0 min; 0.4 mL/min)

99–45% A (2.0–8.0 min; 0.4 mL/min)

45–1% A (8.0–9.0 min; 0.4 mL/min)

1% A (9.0–9.1 min; 0.4–0.6 mL/min)

1% A (9.1–11.0 min; 0.6 mL/min)

1–99% A (11.0–11.1 min; 0.6 mL/min)

99% A (11.1–17.0 min; 0.6 mL/min)

99% A (17.0–17.1 min; 0.6–0.4 mL/min)

99% A (17.1–21.0 min;0.4 mL/min)

Characteristic chromatograms with detection in positive and negative polarity mode are illustrated in Fig. 2.

MS Conditions

For RP-lipid profiling, the mass range is set between 50 to 2000 m/z, while for HILIC between 50 to 1200 m/z. The MS^E mode (continuum mode) is used for MS acquisition. Three functions (parallel acquisition channels) are employed for acquisition: function 1, low collision energy; function 2, high collision energy; and function 3, lock mass acquisition channel. For both low and high collision energy functions, survey scan time is set to 0.2 s. For high collision energy acquisition, collision energy is ramped for low masses 20–40 V and linearly increased up to a 30–50 V ramp, for high masses. Leucine enkephalin (2 ng/µL, 50% ACN, 0.1% FA) is used for lock mass correction. Lock mass data were collected every 30 s for 0.2 s. Other MS parameters were set as follows: cone voltage 30 V, capillary voltage 2 KV, source temperature 120 °C, desolvation temperature 550 °C, and desolvation gas 900 L/h.

UPLC-MS Sample Analysis Setup

Pre-run steps:

1. Perform an initial conditioning of the system by running the gradient either without sample injection or with an injection of a pooled sample or other type of complex mixture (e.g., animal tissue extracts or authentic standards mixture). Inject 10–20 times (*see* **Notes 24** and **25**).

2. Run a test sample. Use this to assess adduct formation, mass accuracy, chromatographic peak width/resolution, and retention time shifting. A standards mix or a biological sample of known metabolite composition (as a more cost-efficient approach), can be used. For tissue types of unknown consistency, a range of commonly detected compounds, which can be used for this task, can be found in Table 1. A sample with pre-determined compounds of known concentration/intensity can also be used to assess that the instrument/method is functioning at a satisfactory sensitivity level (*see* **Note 26**).

3. Second round of conditioning is performed using QC pooled sample. Inject three times. Assess instrument stability and intensity. If signal intensity is low, more sample volume can be injected. *See* next step.

4. Inject 1:2 dilution and assess intensity response to dilution. Reduce injection volume so that signal intensity is decreased

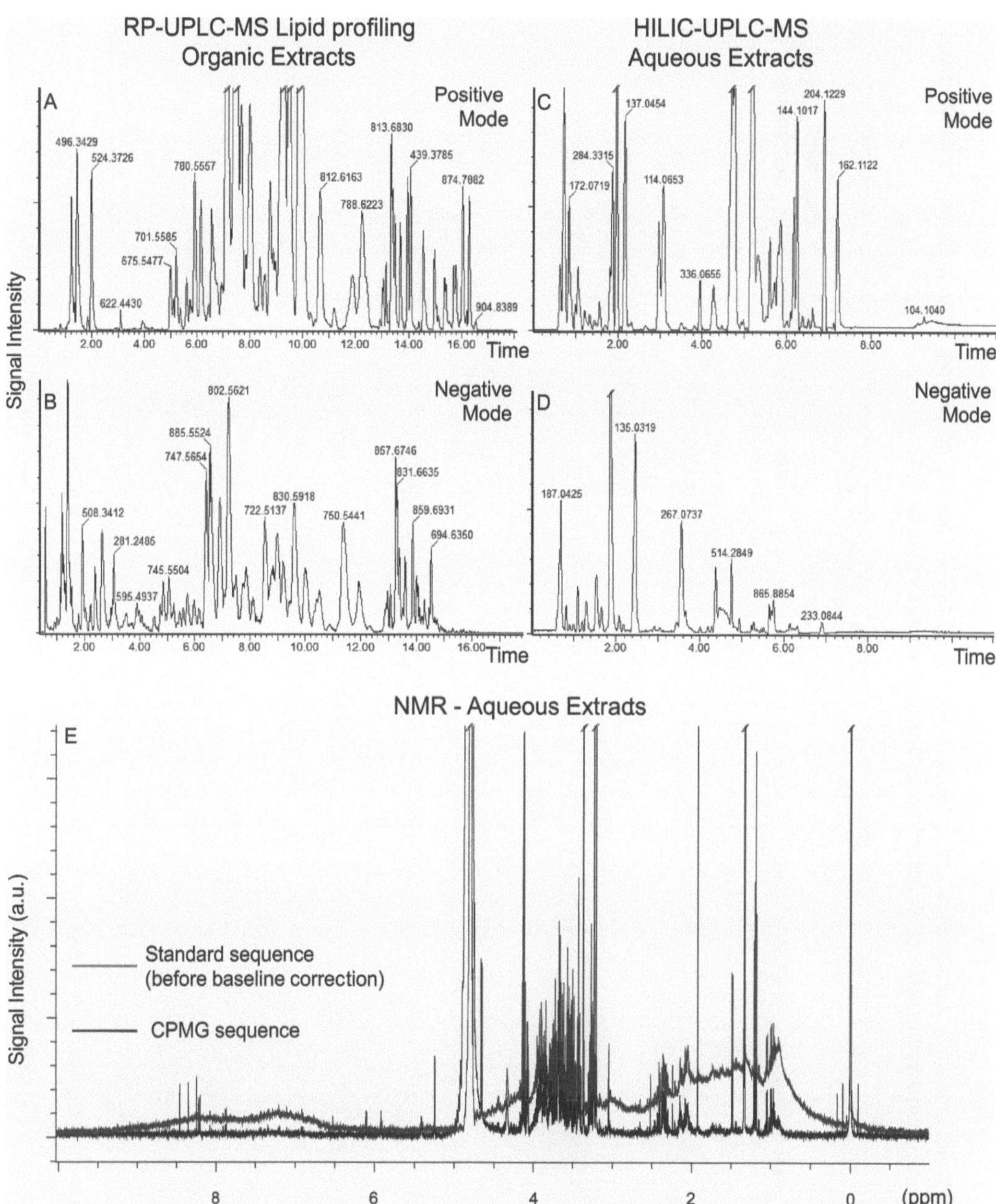

Fig. 2 Characteristic chromatograms/spectra from RP-UPLC-MS analysis of organic extracts (lipid profiling) in (**a**) positive and (**b**) negative modes, HILIC-UPLC-MS analysis of aqueous extracts in (**c**) positive and (**d**) negative modes, and NMR spectra of aqueous extracts acquired using the standard sequence (gray line; prior to baseline correction) and CPMG sequence (black line)

along with the corresponding dilution, and repeat this step. Further dilute the samples if required (*see* **Note 27**).

5. Inject a pooled sample and selected samples from each group or pooled samples of each group, using a data-dependent

Table 1

A range of compounds that are regularly detected in tissue samples, using the described UPLC-MS methods. Signals from these compounds can be used to assess adduct formation, mass accuracy, chromatographic peak width/resolution, and retention time shifting

		Mol formula	Ret time[a] (min)	Mol mass – positive mode (most frequently occurring adduct)	Mol mass – negative mode (most frequently occurring adduct)	Peak width at half max (min)
HILIC	L-carnitine	$C_7H_{15}NO_3$	7.32	162.1130 $[M+H]^+$	–	0.043
	Adenosine	$C_{10}H_{13}N_5O_4$	2.09	268.1046 $[M+H]^+$	266.0895 $[M\text{-}H]-$, 302.0656 $[M+Cl]^-$	0.053
	Creatinine	$C_4H_7N_3O$	3.14	114.0667 $[M+H]^+$	112.0511 $[M–H]^-$	0.062
	Taurine	$C_2H_7NO_3S$	5.80	126.0225 $[M+H]^+$	124.0068 $[M–H]^-$	0.037
	L-leucine[b]	$C_6H_{13}NO_2$	5.83	176.0663 $[M+2Na–H]^+$	130.0868 $[M–H]^-$	0.052
	L-isoleucine[b]	$C_6H_{13}NO_3$	5.96	176.0663 $[M+2Na–H]^+$	130.0868 $[M–H]^-$	0.035
	Glycerophosphocholine	$C_8H_{20}NO_6P$	8.32	258.1106 $[M+H]^+$	242.0793 $[M–CH_3]^-$	0.035
Lipid profiling	Stearic acid	$C_{18}H_{36}O_2$	3.90	–	283.2637 $[M–H]^-$	0.045
	SM(d18:2/16:0)	$C_{39}H_{77}N_2O_6P$	5.20	701.5598 $[M+H]^+$	745.5496 $[M+FA–H]^-$	0.065
	TG(16:0/18:1/18:1)	$C_{55}H_{102}O_6$	16.29	876.8020 $[M+NH_4]^+$	–	0.061
	CE(18:2)	$C_{45}H_{76}O_2$	16.16	666.6189 $[M+NH_4]^+$	–	0.052
	LysoPC(16:0)[c]	$C_{24}H_{50}NO_7P$	1.40	496.3403 $[M+H]^+$	540.3301 $[M+FA–H]^-$	0.048
	LysoPC(0:0/16:0)[c]	$C_{24}H_{50}NO_7P$	1.31	496.3403 $[M+H]^+$	540.3301 $[M+FA–H]^-$	0.048
	PI(18:0/20:4)	$C_{47}H_{83}O_{13}P$	6.56	–	885.5493 $[M–H]^-$	0.078

CE Cholesteryl ester, *FA* formic acid, *SM* sphingomyelin, *lysoPC* lysophosphatidylcholine, *PI* phosphatidylinositol, *TG* triacylglycerol

[a]RT may slightly vary depending on instrument setup

[b]L-leucine and L-isoleucine can be used for ensuring adequate chromatographic resolution in the HILIC method. These two peaks should be fully resolved

[c]LysoPC(16:0) and lysoPC(0:0/16:0) can be used for ensuring adequate chromatographic resolution in the lipid profiling method. These two peaks should be resolved at least at 50%

acquisition (DDA) MS method. Increase the injection volume, if necessary, in order to obtain tandem-MS spectra with higher intensity.

6. Inject a pooled sample and selected samples from each group or pooled samples of each group, using a modified (increase survey scan time to 0.5 s) MS^E method. Increase the injection volume, if required, to obtain tandem-MS spectra with higher intensity.

7. Third round of conditioning using pooled QC sample. Inject five times. Assess instrument stability by comparing repeated injections from this step.

8. Inject reconstitution blank (reconstitution solvent mixture) and assess carry-over. Inject three times.

9. Inject extraction blanks (×1 each) followed by dilution series from low to high concentration (at least ×3 each).

10. Final round of conditioning using pooled QC samples. Inject three times.

Within run:

1. Randomize samples.

2. Include a QC pooled sample injection at least every ten (or less) samples (*see* **Note 28**).

Post-run:

1. Inject the pooled sample dilution series from high to low (at least four times for each dilution of the series) (*see* **Note 29**).

2. Inject extraction blanks (inject each sample once).

3. Inject pooled QC sample three times.

4. Inject the reconstitution blank three times (reconstitution solvent mixture) and assess carry-over.

3.4.3 Data Deconvolution

1. Centroid the UPLC-MS files using MassLynx.

2. Convert centroided files to the NetCDF file format using DataBridge, and use only the first function for further analysis. Alternatively, use MSConvert (ProteoWizard 3.0.11567) to convert to mzXML file format (no further modifications to the succeeding workflow are required).

3. Use the XCMS software to apply:

    (a) Chromatographic peak picking (centWave algorithm).

    (b) Retention time (RT) correction (obiwarp algorithm).

    (c) Peak grouping (density algorithm).

    (d) Fill peaks that resulted in zero intensity, with the intensity of the regional background (fillPeaks algorithm).

    (e) Export data matrix.

3.4.4 Data Normalization

Normalize the data using total area or median fold change normalization (*see* **Note 30**). There are cases where normalization might not be necessary.

3.4.5 Statistical Analysis and Quality Control

1. Import data in SIMCA along with demographic data.

2. Transform and scale the data. For UPLC-MS data a logarithmic transformation followed by scaling is usually recommended. The exploration of various transformation and scaling options according to methodology, instrumentation, and acquired data should be considered.

3. Use OPLS-DA to remove features attributed to the procedure or contamination, by comparing samples against blanks, as previously described [7]. Typically, a cutoff of P(corr) > 0.5 should be used in order to exclude the features attributed to contaminants.

4. Use PCA to evaluate QC samples repeatability. Remove contamination of the QC samples, if present, using PCA or OPLS-DA (Fig. 3). If the latter method is used, a cutoff of P(corr) > 0.8 is typically selected. Remove only features that are exclusively present in the QC samples (*see* **Note 31**).

5. Use PCA for QC dilution feature assessment. Remove features that do not respond to dilutions as previously described [7]. It is recommended to only remove features that elute with the solvent front, or correspond to lipid signals in areas of high coelution of lipids of the same class in HILIC (these areas can be found in previously described in-house databases of annotated lipid species of the HILIC method) [7]. The QC dilution series can also be retrospectively employed to further assess the quality of the discovered putative markers of the studied disease or intervention.

6. Proceed to further statistical analyses and group comparisons.

3.4.6 Metabolite Structure Assignment

Perform metabolite identification as follows:

1. Perform accurate mass database search to obtain possible hits.

2. Match collision-induced dissociation fragments or spectrum to spectra registered in databases.

3. Match authentic standard RT and mass-to-charge ratio to the unknown compound, acquired under identical experimental conditions (*see* **Note 32**).

4. Match an authentic standard MS/MS spectrum to the unknown compound, acquired under identical experimental conditions (*see* **Note 32**).

5. Spike sample(s) with authentic standards if necessary.

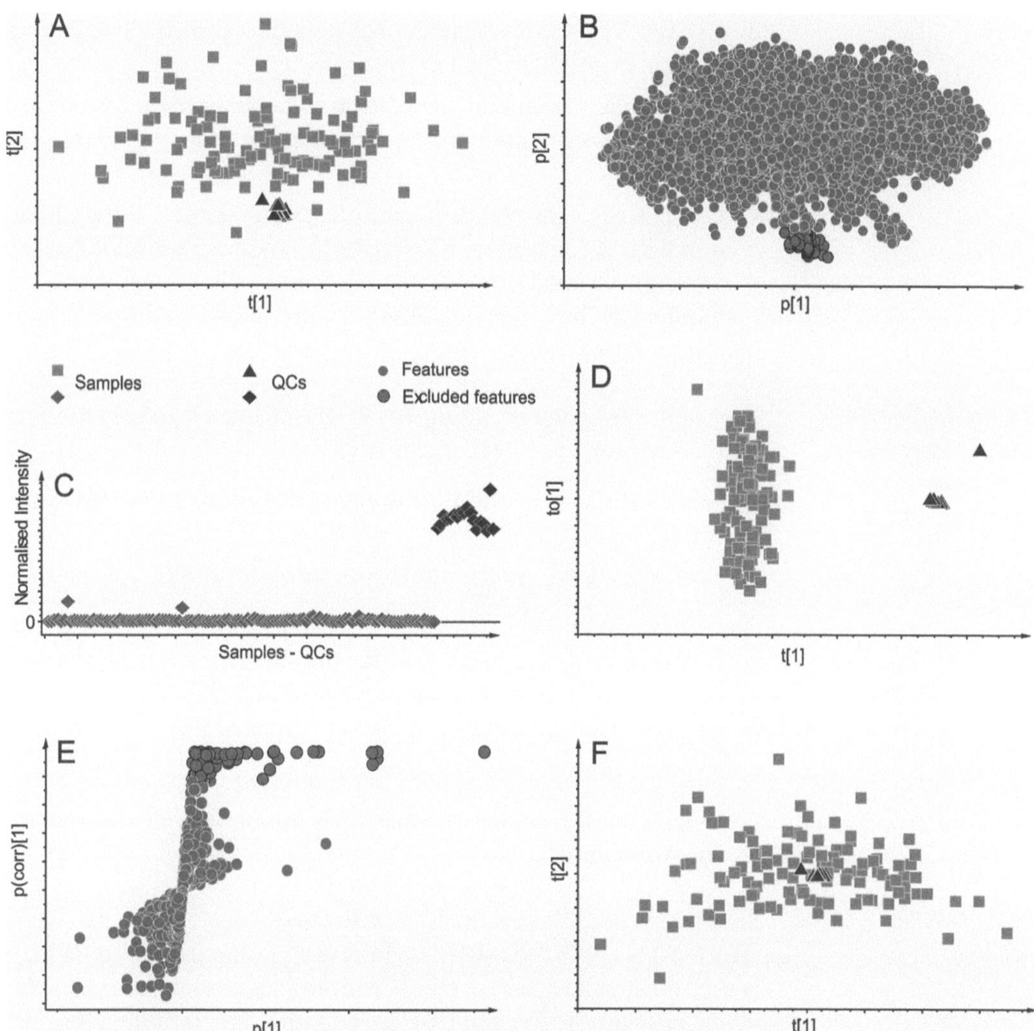

Fig. 3 The quality control (QC) samples may be containing additional contamination, as compared to samples, due to differences between sample/QC preparation. A simple procedure can be followed to remove the features associated with this contamination using multivariate statistics. (**a**) The PCA scores plot demonstrates a contamination in the QC samples, which is not present in the individual samples. (**b**) The loadings plot can assist in pinpointing the contaminating features. (**c**) Feature intensity per sample can help identify features that are only present in QC samples and constitute a contaminant. (**d**) OPLS-DA scores plot. OPLS-DA can be employed as an alternative method for identifying and removing contaminants. (**e**) The S-plot can be used to visualize the features for exclusion. (**f**) The PCA scores plot after features corresponding to contaminants have been removed, showing the QC samples clustering in the center of the individual samples

Additional steps that could aid metabolite structure assignment:

1. Use the isotopic pattern for deciphering elemental composition. This can be done using manual approaches or software (Elemental Composition version 4.0, MassLynx, Waters).

2. Use the RT to assess whether the possible candidate compound structures are plausible (*see* **Note 33**).

3. Use in silico fragment generation (MassFragment, Waters), if corresponding tandem-MS spectra are not available in databases.

4. Use pairwise correlations (*see* Subheading 3.6.2), in combination with RT proximity. Correlations can prove useful for determining structurally related features (isotopes, adducts, fragments) but also metabolites with close biological/(bio) chemical proximity.

3.5 NMR Analysis and Data Treatment

3.5.1 Reconstitution for Aqueous Extracts

Two aliquots (corresponding to 50 mg of tissue sample) are recommended for the NMR acquisition:

1. Add 700 μL of sodium phosphate buffer into the dried tissue extract.

2. Vortex for 1 min followed by sonication for 5 min.

3. Centrifuge for 20 min at 20,000 × g at 4 °C.

4. If two aliquots are used, transfer the supernatant from the first tube to the second, and repeat **steps 2** and **3**.

5. Transfer 600 μL of the supernatant into an NMR tube.

6. Perform the same procedure for extraction blanks.

7. Load a small number of tubes with the buffer only, to test for contamination.

3.5.2 NMR Data Acquisition

Standard one-dimensional NMR sequence [recycle delay (RD)–90°–t_1–90°–t_m–90°–acquire free induction decay (FID)] is used. If standard 1D spectra of the samples contain broad signals which raise up the spectral baseline and interfere with the visualization of the signals derived from the small molecules, the Carr-Purcell-Meiboom-Gill (CPMG) sequence [RD–90°–(τ–180°–τ) n–FID] can also be employed for acquisition (Fig. 2). A total of 512 scans are accumulated into 64 k data points, with a spectral width of 20 ppm. Detailed NMR acquisition parameters can be found in Beckonert et al. [15].

3.5.3 Data Deconvolution and Processing

The following steps are performed in MATLAB using in-house developed scripts:

1. Perform phasing, baseline correction, and calibration using the TSP signal.

2. Remove the TSP and water signals from the spectra.

3. Normalize if necessary. Normalize the data using total area or median fold change normalization (*see* **Note 30**).

4. Align spectra if necessary.

3.5.4 Statistical Analysis and Quality Control	1. Import data in SIMCA along with demographic/external data.

1. Import data in SIMCA along with demographic/external data.

2. Scale the data. Scaling to unit variance or Pareto can be used. Logarithmic transformation prior to scaling can also be employed.

3. Use OPLS-DA to remove features attributed to procedure/solvent contamination, by comparing samples against blanks, as previously described [7]. Typically, a cutoff of P(corr) > 0.5 should be used in order to exclude features attributed to contaminants.

4. Use PCA to evaluate QC samples repeatability. Remove contamination of the QC samples if present using PCA or OPLS-DA. If the latter method is used, a cutoff of P(corr) > 0.8 is typically selected. Remove only features that are exclusively present in the QC samples.

5. Proceed to further statistical analyses and group comparisons.

3.5.5 Metabolite Structure Assignment

Metabolite identification is performed as follows:

1. Perform statistical total correlation spectroscopy (STOCSY) on the signals of interest to search for highly correlated signals, which could be deriving from the same molecule.

2. Check 1D NMR spectra and J-resolved spectra for signal multiplicity.

3. Match the signal chemical shifts and multiplicities to the published literature [16–18], databases such as AMIX (Bruker) and Human Metabolome Database (HMDB), and/or software such as the Chenomx Profiler (Chenomx Inc.).

4. Use 2D NMR spectra, such as ^{1}H-^{1}H COSY (correlation spectroscopy), ^{1}H-^{1}H TOCSY (total correlation spectroscopy), ^{1}H-^{13}C HSQC (heteronuclear single-quantum correlation) and ^{1}H-^{13}C HMBC (heteronuclear multiple bond correlation), to identify the connectivity of the signals.

5. Use hyphenated techniques such as LC-NMR, LC-SPE-NMR, or LC-SPE-NMR/MS to reduce sample complexity, increase the concentrations of the molecules of interest and gain both MS and NMR structural data for metabolite identification.

6. Spike in the samples with the authentic standards of putatively assigned molecules if required.

3.6 Biological Interpretation

For the processes of correlation network construction and pathway mapping, a table of metabolites that demonstrated statistically significant alterations should be produced, along with statistical power, fold change, and values of each metabolite per sample. It is also recommended to assign to each metabolite its corresponding KEGG ID to facilitate importing in pathway mapping software.

3.6.1 Data Integration

Several approaches can be employed for data integration depending on the biological question and techniques/methods used:

1. Combine two or more data matrices by using multiblock multivariate methods, such as O2PLS and OnPLS [19], where extraction of predictive components can be achieved.

2. Perform correlation analysis (*see* Subheading 3.6.2) to obtain associations between entities from the same or different experiments/datasets. Collectively, these associations could lead to inferences assisting in the biological interpretation of the dysregulation or intervention studied.

3. Use appropriate informatics tools (described in Subheading 3.6.3) to perform metabolite enrichment of altered metabolites detected from the same or different experiments. Additionally, metabolite/gene enrichment can be combined if such data are available.

3.6.2 Correlation Analysis

Construct correlation networks after calculating the Spearman correlation coefficients. Correlation coefficients along with their corresponding entity pairs, which could be paired combinations of metabolites, proteins, genes, demographics etc., can then be transferred to the Cytoscape software, where the correlation network can be assembled. All presented correlations in the network should have a $p < 0.05$ (after multiple testing correction), while the cutoff of the correlation coefficient should be set based on the complexity of the illustrated network. The correlation matrix can be demonstrated either as a matrix table or as a heat map.

3.6.3 Pathway Mapping

Import the list of metabolites/KEGG IDs in the KEGG mapper and/or Ingenuity Pathways Analysis software to map the pathways corresponding to the assigned discriminant metabolites. When available, include data from additional biological levels, such as protein and mRNA, in order to combine metabolite and gene enrichment. This can result in higher confidence and statistical power of results.

4 Notes

1. Rinse thrice with LC-MS-grade water followed by rinsing three times with an organic solvent (an organic solvent used for rinsing should be the same as the organic solvent used in the solution to be prepared).

2. Biological hazard: scalpels are extremely sharp and can easily cut through gloves and skin. Dispose immediately after use, and avoid any effort to clean and reuse them.

3. The amount of beads used should remain consistent between tubes (±5% of bead mass). A time-efficient option is to use a

    constant volume of beads (a utensil can be used in this case). Aim for one or two layers of beads above the curved bottom of the tube.

4. Use three steel beads (2.8 mm) for physically harder tissue. In this case, use reinforced tubes. Avoid using solvent volumes that are less than half the volume of the container to prevent breakage of the tube during bead beating.

5. Eppendorf tubes should have the capability of $\geq 25{,}000$ rcf and should be of high quality to avoid contamination due to the leaching of plasticisers and other impurities.

6. Lyophilisation and nitrogen evaporation can also be used for drying down the extraction solvent mixture.

7. If a specific tissue type is expected to have a heavy bacterial load (e.g., colon), then 0.04% NaN_3 should be added in the buffer as a bacteriostatic. The high toxicity of NaN_3 implies that it should be used only when it is absolutely essential.

8. Single-use scalpels and forceps are required to minimize the risk of contamination.

9. Where tissue dissection can be difficult to perform, the analyst can opt for an approach where the total weight of the sample is determined and the extraction solvent is normalized proportionally to the total tissue mass [7].

10. Extraction solvent volume must be accordingly adjusted when large tissue samples are expected for bead beating, to accommodate the increase in total volume.

11. Avoid using proportions of solvent volume-to-tissue weight lower than 4:1.

12. If aliquots for specific targeted methods are to be included in the workflow, it would be optimal to add appropriate internal standards prior to initiating extraction (if method to be used is known beforehand).

13. Check if the tissue has been fully homogenized. Look for completely pulped tissue with no presence of any structure at the bottom of the bead beating tube. If full homogenization has been achieved, avoid additional cycles. Never exceed four cycles. If full homogenization has not been achieved after 4 cycles, then opt for the steel beads option.

14. Do not include blank samples in the preparation of the pooled samples.

15. Ensure complete solvent evaporation. Residual extraction solvents can alter dissolvation efficiency and analyte concentration. Residual methanol can cause a strong contamination signal in NMR spectroscopy.

16. Evaporation time could vary depending on the extraction solvent used, solvent volume, tissue type, and condition of the vacuum concentrator.

17. There are cases where the presence of specific lipids in high concentrations in the tissue sample would make the organic extract appear more liquid, even after full evaporation of the organic solvents.

18. Randomization for the second round of extraction (aqueous extraction) is optional and may not be possible if samples are forwarded for aqueous extraction before organic extraction has been completed for every sample.

19. If tissue has been fully homogenized from the organic extraction step, apply only two beating cycles during the aqueous extraction. Never exceed four cycles.

20. NMR spectroscopy is a non-destructive technique. Samples after being run by NMR can be evaporated and stored for further analysis.

21. Optimal reconstitution volumes may vary according to tissue type and instrumentation. Adjust the reconstitution volume to inject the minimum volume possible. Use the extraction pooled samples for testing and optimizing the system.

22. Decant and combine enough volume to cover the required QC injections.

23. Dilutions of 1:2, 1:3, 1:4, 1:6, and 1:8 of a pooled sample in the reconstitution solvent mixture are suggested. If the extraction pooled sample is used for the dilutions, then a 2:1 concentration option is available and can be prepared by reconstituting with half the volume.

24. Extracts from an easily accessible animal tissue can be used.

25. The number of injections here may depend on the state of the column. New columns may require more injections to condition.

26. Use annotated metabolites from the literature [7]. Use a range of compounds with different masses and physicochemical properties. A range of suggested compounds that are regularly detected in tissue extracts and can be employed to assess adduct formation, mass accuracy, chromatographic peak width/resolution, and retention time shifting can be found in Table 1.

27. If after reducing the injection volume, the ion signal is maintained at detector saturation levels, consider further dilution of the samples. Consider the overall profile, and avoid making decisions based on the (saturated) signals of a small set of ions. It is acceptable to maintain a small number of peaks at

saturation intensities, if this allows for further signals to rise above noise level and become detectable.

28. If a small number of samples (<60) is being run, include at least seven QCs through the run, as this is the minimum required number of repeated injections for robust statistics such as calculating reproducibility using CV%.

29. Inject repeated injections from high to low concentrations to avoid losing column conditioning. The first injection of each dilution can be discarded if necessary. The described methods have demonstrated minimal carry-over [7].

30. Systematic instrument signal reduction due to source contamination would usually require normalization. Total area normalization would typically resolve this issue. Occasionally, differences in tissue composition (normal and diseased) would require normalization procedures such as median fold change [20].

31. The QC sample contamination may be induced due to minor differences in sample handling, such as multiple injections from the same vial or differences employed to accommodate for the larger volume of the QC sample.

32. The same experimental conditions demand running sample and authentic standard one after the other in the same analysis/run. Changes in mass due to differential adduct formation as compared to prior analyses and minor changes in retention time may occur.

33. The log P value (P: partition coefficient) can be used as a measure of lipophilicity and can provide a significant aid in the task of comparing between the level of partitioning of compounds in the chromatographic system and their subsequent differences in RT.

Acknowledgments

This research was supported by the Royal Society of Chemistry and National Institute for Health Research (NIHR) Biomedical Research Centre (BRC) based at Imperial College Healthcare NHS Trust and Imperial College London. The views expressed are those of the authors and not necessarily those of the NHS, the NIHR, or the Department of Health. MRAU is funded by the Imperial College President's PhD Scholarship and the Stratified Medicine Graduate Training Programme in Systems Medicine and Spectroscopic Profiling (STRATiGRAD).

References

1. Lamour SD, Veselkov KA, Posma JM et al (2015) Metabolic, immune, and gut microbial signals mount a systems response to Leishmania major infection. J Proteome Res 14:318–329. https://doi.org/10.1021/pr5008202

2. Folch J, Lees M, Sloane Stanley GH (1957) A simple method for the isolation and purification of total lipides from animal tissues. J Biol Chem 226:497–509

3. Bligh EG, Dyer WJ (1959) A rapid method of total lipid extraction and purification. Can J Biochem Physiol 37:911–917

4. Geier FM, Want EJ, Leroi AM et al (2011) Cross-platform comparison of Caenorhabditis elegans tissue extraction strategies for comprehensive metabolome coverage. Anal Chem 83:3730–3736. https://doi.org/10.1021/ac2001109

5. Masson P, Spagou K, Nicholson JK et al (2011) Technical and biological variation in UPLC-MS-based untargeted metabolic profiling of liver extracts: application in an experimental toxicity study on galactosamine. Anal Chem 83:1116–1123. https://doi.org/10.1021/ac103011b

6. Anwar MA, Vorkas P, Li JV et al (2015) Optimization of metabolite extraction of human vein tissue for ultra performance liquid chromatography-mass spectrometry and nuclear magnetic resonance-based untargeted metabolic profiling. Analyst 140:7586–7597

7. Vorkas PA, Isaac G, Anwar MA et al (2015) Untargeted UPLC-MS profiling pipeline to expand tissue metabolome coverage: application to cardiovascular disease. Anal Chem 87:4184–4193. https://doi.org/10.1021/ac503775m

8. Vorkas PA, Shalhoub J, Isaac G et al (2015) Metabolic phenotyping of atherosclerotic plaques reveals latent associations between free cholesterol and ceramide metabolism in atherogenesis. J Proteome Res 14:1389–1399. https://doi.org/10.1021/pr5009898

9. Ashrafian H, Li JV, Spagou K et al (2014) Bariatric surgery modulates circulating and cardiac metabolites. J Proteome Res 13:570–580. https://doi.org/10.1021/pr400748f

10. Anwar MA, Vorkas PA, Li J et al (2016) Prolonged mechanical circumferential stretch induces metabolic changes in rat inferior vena cava. Eur J Vasc Endovasc 52:544–552. https://doi.org/10.1016/j.ejvs.2016.07.002

11. Vorkas PA, Shalhoub J, Lewis MR et al (2016) Metabolic phenotypes of carotid atherosclerotic plaques relate to stroke risk: an exploratory study. Eur J Vasc Endovasc 52:5–10. https://doi.org/10.1016/j.ejvs.2016.01.022

12. Anwar MA, Adesina-Georgiadis KN, Spagou K et al (2017) A comprehensive characterisation of the metabolic profile of varicose veins; implications in elaborating plausible cellular pathways for disease pathogenesis. Sci Rep 7:2989. https://doi.org/10.1038/s41598-017-02529-y

13. Shannon P, Markiel A, Ozier O et al (2003) Cytoscape: a software environment for integrated models of biomolecular interaction networks. Genome Res 13:2498–2504. https://doi.org/10.1101/gr.1239303

14. Kanehisa M, Goto S (2000) KEGG: Kyoto encyclopedia of genes and genomes. Nucleic Acids Res 28:27–30

15. Beckonert O, Keun HC, Ebbels TM et al (2007) Metabolic profiling, metabolomic and metabonomic procedures for NMR spectroscopy of urine, plasma, serum and tissue extracts. Nat Protoc 2:2692–2703. https://doi.org/10.1038/nprot.2007.376. nprot.2007.376 [pii]

16. Nicholson JK, Foxall PJ, Spraul M et al (1995) 750 MHz 1H and 1H-13C NMR spectroscopy of human blood plasma. Anal Chem 67:793–811

17. Yap IK, Brown IJ, Chan Q et al (2010) Metabolome-wide association study identifies multiple biomarkers that discriminate north and south Chinese populations at differing risks of cardiovascular disease: INTERMAP study. J Proteome Res 9:6647–6654. https://doi.org/10.1021/pr100798r

18. Saric J, Wang Y, Li J et al (2008) Species variation in the fecal metabolome gives insight into differential gastrointestinal function. J Proteome Res 7:352–360. https://doi.org/10.1021/pr070340k

19. Lofstedt T, Trygg J (2011) OnPLS-a novel multiblock method for the modelling of predictive and orthogonal variation. J Chemom 25:441–455. https://doi.org/10.1002/cem.1388

20. Veselkov KA, Vingara LK, Masson P et al (2011) Optimized preprocessing of ultra-performance liquid chromatography/mass spectrometry urinary metabolic profiles for improved information recovery. Anal Chem 83:5864–5872. https://doi.org/10.1021/ac201065j

Chapter 18

UHPLC-HRMS Analysis for Steroid Profiling in Serum (Steroidomics)

Federico Ponzetto, Julien Boccard, Raul Nicoli, Tiia Kuuranne, Martial Saugy, and Serge Rudaz

Abstract

The extraction and untargeted UHPLC-HRMS analysis of endogenous steroids in serum samples is described in this protocol. The employed full-scan acquisition mode provides the adequate sensitivity to highlight the main endogenous steroids present in blood, including mineralocorticoids, progestogens, and androgens. Technical aspects for both chromatography and mass spectrometry are discussed in detail, together with a proposition of setup for sample sequence and data analysis. Furthermore, general comments are given to help the assessment of data quality and system performance.

Key words Steroids, Extended steroid profiles, Steroidomics, UHPLC-HRMS

1 Introduction

Steroids are structurally related hormones originating from cholesterol synthesized in numerous organs, including the adrenal glands, testis, ovaries, brain, placenta, and adipose tissue [1, 2], which regulate many essential functions such as growth, metabolic rate, sexual functions, and stress response. Because altered regulation of steroid metabolism (due to genetic or environmental factors) is often related to severe pathologies, the large-scale monitoring of steroidogenesis provides crucial information for therapeutic purpose [3]. Furthermore, the persistent misuse of anabolic androgenic steroids, which perturb testosterone metabolism leading to better performance/recovery or health issues, raises the need of such a monitoring also in anti-doping field [4–7].

Steroid analysis for the diagnosis of endocrine diseases was initially achieved by gas chromatography-based (GC) methods in the 1950s and since 1970 was then complemented by high-throughput immunoassays (IA) [8]. However, as IA techniques often revealed problems of cross-reactivity, particularly in the case of steroids, a

Georgios A. Theodoridis et al. (eds.), *Metabolic Profiling: Methods and Protocols*, Methods in Molecular Biology, vol. 1738, https://doi.org/10.1007/978-1-4939-7643-0_18, © Springer Science+Business Media, LLC, part of Springer Nature 2018

standardization of steroid tests employing separation-based techniques, such as chromatographic methods coupled to mass spectrometry, was recommended [9–11]. Coupling mass spectrometry to either liquid chromatography (LC-MS) or gas chromatography (GC-MS) is the current reference approach for measuring steroids; in particular, the latter, even if limited by the need for a derivatization step aiming at improving the volatility of the compounds prior to analysis, constitutes the reference for anti-doping analyses [12]. More recently, untargeted profiling strategies combining the high peak capacity provided by ultrahigh pressure liquid chromatography (UHPLC) and the resolution offered by high-resolution mass spectrometry (HRMS) are representing an appealing alternative for simultaneously measuring hundreds of steroids in complex biological matrices. This steroidomic approach was first described by Sjövall in 2004 and defined as the "characterization and quantification of metabolic profiles of steroids" [13].

In this protocol an untargeted steroidomic analysis on serum samples using a UHPLC-HRMS platform is presented. The overall methodology has been optimized to provide high extraction recoveries for steroids in serum samples as well as good analytical performance for both sensitivity and chromatographic separation, using 101 endogenous steroids as reference standards. The proposed procedure could also be adapted for absolute quantification of steroids of particular interest, but in general the development of a separate assay is preferred for such a purpose.

2   Materials

Ultrapure water (18.2 MΩ, total organic carbon <5 ppb), organic solvents (acetonitrile, ACN, and methanol, MeOH), and formic acid (FA) of HPLC grade or higher purity should be used. The same quality criteria must be applied to analytical standards to obtain the highest available purity. Reference endogenous steroid standards as well as deuterium-labeled internal standards could be purchased on the market from various providers, such as Sigma-Aldrich, Steraloids, and Cerilliant. Charcoal stripped human serum (steroid depleted/negative serum) is purchased from Dunn Labortechnik GmbH (Asbach, Germany).

2.1 Analytes Stock Solutions, Internal Standard, and Quality Control Mixtures

Endogenous steroids and labeled internal standards should be purchased as calibrated stock solutions (1 mg/mL or 100 µg/mL in MeOH or ACN) or as powder. For the latter, stock solutions are prepared by weighing the appropriate amount of powder and dissolving it in MeOH to obtain a final concentration of 1 mg/mL. Stock solutions are finally diluted in MeOH to obtain intermediate solutions at a concentration of 10 µg/mL. Both stock and intermediate solutions are stored at −80 °C and should be thawed only when it is strictly mandatory to limit degradation.

Testosterone-d3,   androsterone-d4, 17α-hydroxyprogesterone-d8, and cortisol-d4 stock solutions are diluted in MeOH to obtain an Internal Standard Mix solution (IS-mix) at a concentration of 3 ng/mL, 10 ng/mL, 5 ng/mL, and 25 ng/mL, respectively. The IS-mix is stored at −20 °C and should be thawed at least 30 min before its utilization during sample preparation.

Quality Control Mix (QC-mix) is prepared by mixing equal aliquots of all serum samples that have to be analyzed in the study (see also Chapter 2). It is important to prepare a sufficient amount of QC-mix aliquots before starting the analyses and store them at −20 °C. For each analytical batch, the required number of QC-mix aliquots should be treated as all the other serum samples and thawed overnight in refrigerator at 4 °C the day before extraction.

2.2 Solvent Preparation for Liquid Chromatography

1. *Mobile Phase A:* In a 1 L solvent bottle, add 1 mL of FA to 1 L of ultrapure H_2O and shake to ensure mixing. This solution is stored at room temperature and expires after 7 days (*see* **Notes 1** and **3**).

2. *Mobile Phase B:* Add 2.5 mL of FA to a 2.5 L bottle of ACN and shake to ensure mixing. This solution is stored at room temperature and expires after 4 months (*see* **Notes 2** *and* **3**).

3. *Weak Wash Solution:* In a 1 L solvent bottle, mix 980 mL of ultrapure H_2O with 20 mL of ACN and shake to ensure mixing. This solution is stored at room temperature and expires after 7 days (*see* **Note 3**).

4. *Strong Wash Solution:* 1 L of pure ACN is used in an appropriate bottle. This solvent is stored at room temperature and expires after 4 months (*see* **Note 3**).

2.3 Instrumentation

Sample preparation is performed using a PRESSURE+ 96 Positive Pressure Manifold (Biotage, Uppsala, Sweden). Analyses are carried out using a Waters Acquity UPLC system (Milford, MA, USA) including a binary solvent manager, a sample manager equipped with an external fixed loop of 20 µL, and a column manager. The UPLC is coupled to a Q Exactive Plus mass spectrometer (Thermo Fisher Scientific, Waltham, MA, USA). Data are acquired and processed using Xcalibur PC software (version 3.1), and the data treatment is performed with Progenesis QI software (Version 2.0, 64-bit, Nonlinear Dynamics, Newcastle upon Tyne, UK).

3   Methods

3.1   Serum Sample Extraction

Supported liquid extraction (SLE) on ISOLUTE® SLE+ 400 µL 96-well plates (Biotage, Uppsala, Sweden) is used to extract steroid hormones from serum samples. The extraction procedure consists of five different steps:

1. For each sample, an aliquot of 200 µL of serum is spiked with 20 µL of the IS-mix, diluted with 200 µL of water, and then agitated for 15 min at 250 rpm.

2. Each well is then loaded with 400 µL of the previously diluted sample, and a positive pressure of 3 psi is applied for 30 s to facilitate sample loading and adsorption.

3. The elution is carried out, after a 5 min waiting period, by adding 700 µL of dichloromethane to each well and applying a pressure of 3 psi for 1 min.

4. Extracts are then collected in collection plates equipped with 1.5 mL glass inserts, evaporated to dryness for approximately 15 min at 40 °C under a gentle stream of nitrogen, and finally reconstituted with 100 µL of a MeOH-H_2O 50:50 (v/v) mixture (reconstitution solvent).

5. After 15 min of gentle shaking (250 rpm), 10 µL of each extract is injected into UHPLC-HRMS system for analyses.

3.2   UHPLC

Liquid chromatography is performed with a Kinetex C_{18} column (150 × 2.1 mm, 1.7 µm; Phenomenex, Torrance, CA, USA) set at 30 °C. The sample injection volume is 10 µL (partial loop mode), and the flow rate is set at 300 µL/min. The mobile phase A is a solution of 0.1% FA in H_2O, and the mobile phase B is 0.1% FA in ACN. The linear gradient starts from 5% of B and increases linearly to 95% over 16.8 min, followed by 2 min of plateau at 95% B; the column is then re-equilibrated for 7 min at initial conditions. The total run time is 25.9 min (*see* **Notes 4** and **5**).

3.3   Mass Spectrometry

HRMS analyses are performed in positive electrospray ionization (ESI) in both full-scan (FS) and data-dependent MS/MS (ddMS²) acquisition modes. Mass calibration (<3 ppm) is performed before each analytical sequence using the Pierce® LTQ Velos ESI Positive Ion Calibration standard mixture (Thermo Fisher Scientific) containing *n*-butylamine, caffeine, MRFA (peptide of Met-Arg-Phe-Ala acetate salt), and Ultramark 1621. The heated ESI source (HESI II) is used with a probe heater temperature of 425 °C. The sheath gas and auxiliary gas pressures are set to 50 and 15 arbitrary units, respectively, and the sweep gas flow is set to 3 arbitrary units. The ion spray voltage is set to 4.5 kV, the capillary temperature to 250 °C, and the S-Lens RF level to 55%. FS mass spectra are acquired in profile mode using a mass resolution of 70,000 (full width at half maximum, FWHM) at m/z 200, with a maximum IT fill time of 125 ms, and the automatic gain control (AGC target) set to $3e^6$. The acquired mass range is from m/z 200 to 600. The $ddMS^2$ acquisition mode is performed using a mass resolution of 17,500 FWHM, with a maximum IT fill time of 64 ms and the AGC target set to $5e^4$. The isolation window is set to 0.4 m/z, and the number of different ions to be fragmented after each FS event (loop count) is set to 5, using stepped normalized collision energy values of 20, 40, and 60. For the data-dependent acquisition, a minimum AGC target of $5e^3$ is set, resulting in an intensity threshold of $7.8e^4$, above which the MS/MS analysis is triggered. For improved selectivity, both apex trigger and dynamic exclusion features are enabled (*see* **Note 6**). All detailed parameters of FS-ddMS^2 experiment are presented in Fig. 1.

3.4   HRMS
Maintenance
and Calibration

Prior to the injection of the extracted serum samples, it is recommended to carry out the cleaning of the ESI source followed by the calibration of the MS system. The MS cleaning procedure involves the cone and the ion transfer tube that are removed from the MS instrument after cooling the ESI source temperature to room temperature. The instrument is then switched on standby mode, and the two parts of the ion source are subjected to three washing steps of 10 min sonication each in three different solvents:

1. 10% FA in ultrapure H_2O.

2. Ultrapure H_2O.

3. HPLC grade MeOH.

Once the third step is completed, the cone and the ion transfer tube are dried under a steam of argon and then assembled again on the instrument (*see* **Note 7**). Then, the calibration MS tune file (sheath gas flow rate set to 3 arbitrary units, both auxiliary and sweep gas flow set to 0; ion spray voltage of 3.5 kV and capillary temperature of 320 °C with the S-Lens RF level at 50%) is loaded and the ESI source connected to the syringe filled up with the Positive Ion Calibration standard mixture. The flow rate is set at 5 μL/min, and the system should be stabilized for 10 min; the calibration is performed only when the TIC variation is constantly lower than 10%.

3.5   In-House
Endogenous Steroids
Database

When setting up a steroidomic study, it is important not only to develop a suitable analytical procedure but also to measure adequate reference material, with which data of the untargeted acquisitions could be matched for the identification of most interesting compounds. This protocol is based on the use of an in-house database, which consists of up to 101 endogenous steroids, including

General	
Runtime	1 to 17 min
Polarity	positive
In-source CID	0.0 eV
Default charge state	1
Inclusion	-
Exclusion	-
Tags	-
Full MS	
Microscans	1
Resolution	70,000
AGC target	3e6
Maximum IT	125 ms
Number of scan range	1
Scan range	200 to 600 m/z
Spectrum data type	Profile
dd-MS^2 / dd-SIM	
Microscans	1
Resolution	17,500
AGC target	5e4
Maximum IT	64 ms
Loop count	5
MSX count	1
TopN	5
Isolation window	0.4 m/z
Isolation offset	0.0 m/z
Fixed first mass	-
NCE / stepped NCE	20, 40, 60
Spectrum data type	Profile
dd Settings	
Minimum AGC target	5.00e3
Intensity threshold	7.8e4
Apex trigger	2 to 6 s
Charge exclusion	-
Peptide match	-
Exclude isotopes	-
Dynamic exclusion	6.0 s

Fig. 1 Full-scan and data-dependent MS/MS experiment parameters

major androgens, progestogens, mineralocorticoids, and their phase I metabolites, but the number of standards strongly depends on the foreseen application. For this purpose, working solutions of each analyte (*see* Table 1) at a concentration of 100 ng/mL are prepared in the reconstitution solvent and injected in the UHPLC-HRMS system using the analytical method described above. From these injections, information on retention time and ionization properties of all the target endogenous steroids has been retrieved and summarised in Table 1.

3.6 Performance Monitoring

To monitor sample preparation and UHPLC-HRMS system performance, two different strategies based on IS-mix and QC-mix are considered:

1. Regarding extraction, 20 µL of the IS-mix containing the four deuterium-labeled internal standards of different endogenous steroids (testosterone-d3 and androsterone-d4, androgens; 17α-hydroxyprogesterone-d8, progestogen; cortisol-d4, mineralocorticoid) are spiked in all samples. A noticeable reduced peak area of IS-mix compounds could be the signal of a problem in the sample preparation suggesting the re-extraction of a sample.

2. QC-mix is used in steroidomic studies for conditioning the analytical platform at the beginning of the sample sequence. Furthermore, the injection of QC-mix at regular intervals during the analytical sequence allows for the assessment of analytical variability and instrument performance (*see* **Notes 8** and **9**).

Chromatograms of selected identified compounds are presented in Fig. 2.

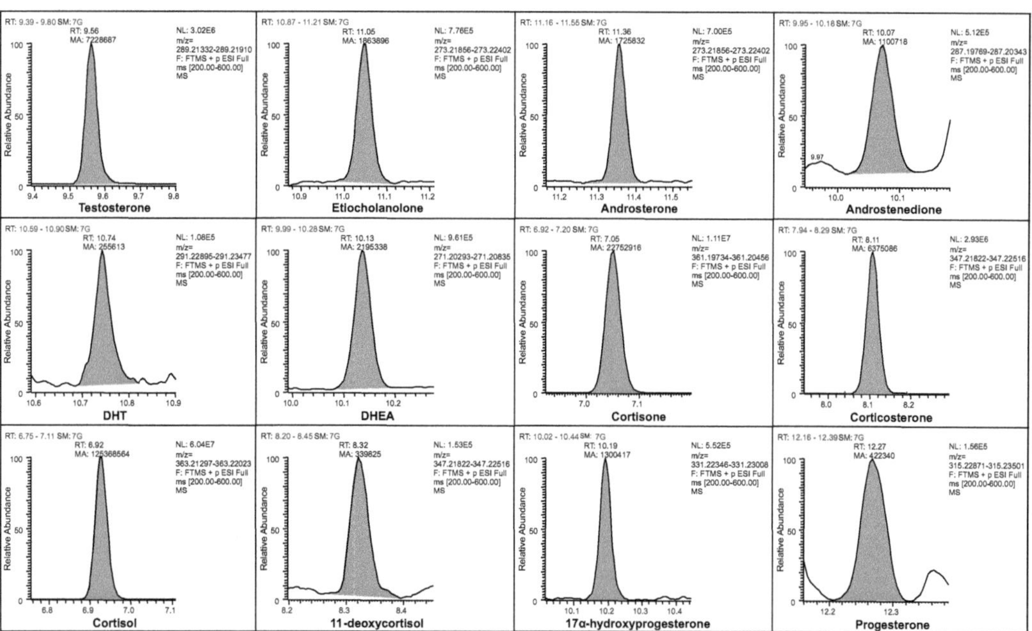

Fig. 2 Chromatograms of 12 selected endogenous steroids extracted from a QC-mix sample

Table 1
Chemical formula, retention time, and most abundant ion m/z value of the 101 endogenous steroids of the in-house steroid database

Compound name	Chemical formula	Exact mass [Da]	Retention time [min]	Most abundant ion [m/z]			
				$[M+H]^+$	$[M+Na]^+$	$[M-H_2O]^+$	$[M-2H_2O]^+$
11α-Hydroxyprogesterone	$C_{21}H_{30}O_3$	330.21950	8.82	331.22677			
11β-Hydroxyandrostenedione	$C_{19}H_{26}O_3$	302.18820	8.24	303.19547			
11β-Hydroxyandrosterone	$C_{19}H_{30}O_3$	306.21950	8.69				271.20564
11β-Hydroxyepiandrosterone	$C_{19}H_{30}O_3$	306.21950	8.22				271.20564
11β-Hydroxyetiocholanolone	$C_{19}H_{30}O_3$	306.21950	8.56				271.20564
11β-Hydroxyprogesterone	$C_{21}H_{30}O_3$	330.21950	9.89	331.22677			
11β-Hydroxytestosterone	$C_{19}H_{28}O_3$	304.20385	7.56	305.21112			
11-Dehydrocorticosterone	$C_{21}H_{28}O_4$	344.19876	7.87	345.20604			
11-Dehydrotetrahydrocorticosterone	$C_{21}H_{32}O_4$	348.23006	8.43			331.22677	
11-Deoxycorticosterone	$C_{21}H_{30}O_3$	330.21950	9.79	331.22677			
11-Deoxycortisol	$C_{21}H_{30}O_4$	346.21441	8.36	347.22169			
11-Ketoetiocholanolone	$C_{19}H_{28}O_3$	304.20385	8.92			287.20056	
11-Oxoandrosterone	$C_{19}H_{28}O_3$	304.20385	8.97	305.21112			
15β-Hydroxytestosterone	$C_{19}H_{28}O_3$	304.20385	6.28	305.21112			
16α-Hydroxyandrostenediol	$C_{19}H_{30}O_3$	306.21950	6.51				271.20564
16α-Hydroxyandrostenedione	$C_{19}H_{26}O_3$	302.18820	7.74	303.19547			
16α-Hydroxyandrosterone	$C_{19}H_{30}O_3$	306.21950	9.10			289.21621	
16α-Hydroxydehydroepiandrosterone	$C_{19}H_{28}O_3$	304.20385	7.62			287.20056	

16α-Hydroxyetiocholanolone	$C_{19}H_{30}O_3$	306.21950	9.02		289.21621
16α-Hydroxyprogesterone	$C_{21}H_{30}O_3$	330.21950	8.42	331.22677	
16α-Hydroxytestosterone	$C_{19}H_{28}O_3$	304.20385	6.85	305.21112	
16β-Hydroxydehydroepiandrosterone	$C_{19}H_{28}O_3$	304.20385	7.29		287.20056
16β-Hydroxytestosterone	$C_{19}H_{28}O_3$	304.20385	7.63	305.21112	
17α,20α-Dihydroxyprogesterone	$C_{21}H_{32}O_3$	332.23515	9.07	333.24242	
17α-Hydroxypregnenolone	$C_{21}H_{32}O_3$	332.23515	9.96		315.23186
17α-Hydroxyprogesterone	$C_{21}H_{30}O_3$	330.21950	10.25	331.22677	
19-Hydroxyandrostenedione	$C_{19}H_{26}O_3$	302.18820	7.07	303.19547	
19-Hydroxytestosterone	$C_{19}H_{28}O_3$	304.20385	6.62	305.21112	
20α-Cortolone	$C_{21}H_{34}O_5$	366.24063	6.99		331.22677
20α-Dihydrocortisone	$C_{21}H_{30}O_5$	362.20933	6.27	363.21660	
20α-Dihydroprogesterone	$C_{21}H_{32}O_2$	316.24023	11.21	317.24751	
20β-Cortol	$C_{21}H_{36}O_5$	368.25628	6.82		333.24242
20β-Cortolone	$C_{21}H_{34}O_5$	366.24063	7.11		331.22677
21-Deoxycortisol	$C_{21}H_{30}O_4$	346.21441	8.15	347.22169	
21-Hydroxypregnenolone	$C_{21}H_{32}O_3$	332.23515	9.73		315.23186

(continued)

Table 1
(continued)

Compound name	Chemical formula	Exact mass [Da]	Retention time [min]	Most abundant ion [m/z]			
				$[M+H]^+$	$[M+Na]^+$	$[M-H_2O]^+$	$[M-2H_2O]^+$
2α-Hydroxytestosterone	$C_{19}H_{28}O_3$	304.20385	7.87	305.21112			
2β-Hydroxytestosterone	$C_{19}H_{28}O_3$	304.20385	7.90	305.21112			
3α,21-Dihydroxy-5α-pregnane-11,20-dione	$C_{21}H_{32}O_4$	348.23006	8.43	349.23734			
3α,5α-Tetrahydrocorticosterone	$C_{21}H_{34}O_4$	350.24571	8.46				315.23186
3α,5α-Tetrahydrocortisol	$C_{21}H_{34}O_5$	366.24063	7.41				331.22677
3α,5β-Tetrahydroaldosterone	$C_{21}H_{32}O_5$	364.22498	6.54		387.21420		
3α,5β-Tetrahydrocorticosterone	$C_{21}H_{34}O_4$	350.24571	8.34				315.23186
3α,5β-Tetrahydrocortisol	$C_{21}H_{34}O_5$	366.24063	7.45				331.22677
3α,5β-Tetrahydrocortisone	$C_{21}H_{32}O_5$	364.22498	7.77			347.22169	
5α,20α-Tetrahydroprogesterone	$C_{21}H_{34}O_2$	318.25588	12.47			301.25259	
5α-Androstane-3α,17β-diol	$C_{19}H_{32}O_2$	292.24023	10.61				257.22638
5α-Androstan-3β-ol-7,17-dione	$C_{19}H_{28}O_3$	304.20385	7.36			287.20056	
5α-Androstane-3β,12β,15α-triol	$C_{19}H_{32}O_3$	308.23515	5.69				273.22129
5α-Androstane-3β,7α,16β-triol	$C_{19}H_{32}O_3$	308.23515	6.73				273.22129
5α-Androstanedione	$C_{19}H_{28}O_2$	288.20893	11.35	289.21621			
5α-Androstane-3β,17β-diol	$C_{19}H_{32}O_2$	292.24023	9.97				275.23694

5α-Dihydro-11-deoxycorticosterone	$C_{21}H_{32}O_3$	332.23515	10.98	333.24242			
5α-Dihydrocorticosterone	$C_{21}H_{32}O_4$	348.23006	8.98	349.23734			
5α-Dihydrocortisol	$C_{21}H_{32}O_5$	364.22498	7.54	365.23225			
5α-Dihydrocortisone	$C_{21}H_{30}O_5$	362.20933	7.58	363.21660			
5α-Dihydroprogesterone	$C_{21}H_{32}O_2$	316.24023	13.47	317.24751			
5-Androsten-3β,17β-diol-16-one	$C_{19}H_{28}O_3$	304.20385	7.31		287.20056		
5β-Androstane-3α,17β-diol	$C_{19}H_{32}O_2$	292.24023	10.36			275.23694	
5β-Androstan-11α-ol-3,17-dione	$C_{19}H_{28}O_3$	304.20385	8.53				269.18999
5β-Androstan-11β-ol-3,17-dione	$C_{19}H_{28}O_3$	304.20385	9.13	305.21112			
5β-Androstane-3α,17α-diol	$C_{19}H_{32}O_2$	292.24023	11.57				257.22638
5β-Androstane-3β,17α-diol	$C_{19}H_{32}O_2$	292.24023	10.74				257.22638
5β-Androstanedione	$C_{19}H_{28}O_2$	288.20893	11.45	289.21621			
5β-Androstane-3β,17β-diol	$C_{19}H_{32}O_2$	292.24023	10.08			275.23694	
5β-Dihydro-11-dehydrocorticosterone	$C_{21}H_{30}O_4$	346.21441	8.87	347.22169			
5β-Dihydrocorticosterone	$C_{21}H_{32}O_4$	348.23006	9.02	349.23734			
5β-Dihydrocortisol	$C_{21}H_{32}O_5$	364.22498	7.86		347.22169		
5β-Dihydrocortisone	$C_{21}H_{30}O_5$	362.20933	8.10	363.21660			
5β-Dihydroprogesterone	$C_{21}H_{32}O_2$	316.24023	13.38	317.24751			

(continued)

Table 1
(continued)

Compound name	Chemical formula	Exact mass [Da]	Retention time [min]	Most abundant ion [m/z]			
				$[M+H]^+$	$[M+Na]^+$	$[M-H_2O]^+$	$[M-2H_2O]^+$
5β-Dihydrotestosterone	$C_{19}H_{30}O_2$	290.22458	10.85	291.23186			
6α-Hydroxytestosterone	$C_{19}H_{28}O_3$	304.20385	6.36	305.21112			
6β-Hydroxytestosterone	$C_{19}H_{28}O_3$	304.20385	6.69	305.21112			
7α-Hydroxydehydroepiandrosterone	$C_{19}H_{28}O_3$	304.20385	7.00			287.20056	
7α-Hydroxypregnenolone	$C_{21}H_{32}O_3$	332.23515	8.28			315.23186	
7α-Hydroxytestosterone	$C_{19}H_{28}O_3$	304.20385	6.43	305.21112			
7β-Hydroxydehydroepiandrosterone	$C_{19}H_{28}O_3$	304.20385	6.63			287.20056	
Adrenosterone	$C_{19}H_{24}O_3$	300.17255	8.38	301.17982			
Aldosterone	$C_{21}H_{28}O_5$	360.19368	6.48	361.20095			
Allotetrahydrocortisol	$C_{21}H_{34}O_5$	366.24063	7.41				331.22677
Allotetrahydrodesoxycorticosterone	$C_{21}H_{34}O_3$	334.25080	10.65			317.24751	
Androst-5-ene-3β,16α,17α-triol	$C_{19}H_{30}O_3$	306.21950	7.92				271.20564
Androstan-3β,5α,6β-triol	$C_{19}H_{32}O_3$	308.23515	9.92				273.22129
Androst-5-ene-3β,16β,17α-triol	$C_{19}H_{30}O_3$	306.21950	6.61				271.20564
Androstenedione	$C_{19}H_{26}O_2$	286.19328	10.12	287.20056			
Androsterone	$C_{19}H_{30}O_2$	290.22458	11.41			273.22129	
Corticosterone	$C_{21}H_{30}O_4$	346.21441	8.16	347.22169			
Cortisol	$C_{21}H_{30}O_5$	362.20933	6.97	363.21660			

Cortisone	$C_{21}H_{28}O_5$	360.19368	7.10	361.20095		
Dehydroandrosterone	$C_{19}H_{28}O_2$	288.20893	10.55		271.20564	
Dehydroepiandrosterone	$C_{19}H_{28}O_2$	288.20893	10.18		271.20564	
Dihydrotestosterone	$C_{19}H_{30}O_2$	290.22458	10.79	291.23186		
Epiandrosterone	$C_{19}H_{30}O_2$	290.22458	10.63		273.22129	
Epitestosterone	$C_{19}H_{28}O_2$	288.20893	10.20	289.21621		
Etiocholanolone	$C_{19}H_{30}O_2$	290.22458	11.12		273.22129	
Pregnanediol	$C_{21}H_{36}O_2$	320.27153	12.10			285.25768
Pregnanetriol	$C_{21}H_{36}O_3$	336.26645	10.38			301.25259
Pregnanolone	$C_{21}H_{34}O_2$	318.25588	12.86		301.25259	
Pregnenolone	$C_{21}H_{32}O_2$	316.24023	12.25		299.23694	
Progesterone	$C_{21}H_{30}O_2$	314.22458	12.21	315.23186		
Testosterone	$C_{19}H_{28}O_2$	288.20893	9.59	289.21621		
Tetrahydro-11-deoxycortisol	$C_{21}H_{34}O_4$	350.24571	9.52			315.23186

<table>
<tr><td>

3.7 Analytical Sequence

</td><td>

The analytical sequence should be constructed carefully with the aim of simultaneously providing the best compromise between a satisfying data acquisition and a clear measure of the data quality and system performance. Solvent blank, negative serum samples, conditioning samples, and QC-mix samples are added to the steroidomic study serum sample injections helping the assessment of the analysis quality. The proposal of a steroidomic analytical sequence is described below:

</td></tr>
</table>

1. A solvent blank injection at the beginning of the analytical sequence helps the identification of background noise related to the reconstitution solvent. If more than one solvent blank injection is performed, it is also possible to evaluate the presence of any carry over effect.

2. Before starting the conditioning process, a minimum of two injections of negative serum samples should be performed. These injections are useful in evaluating the status of the analytical system, in particular in regard to the chromatographic column. Indeed, the injection of negative serum samples could help the assessment of any carry over of endogenous steroids related to previous analyses, which is difficult to detect when injecting the solvent blank.

3. After the negative serum sample injections, a minimum of five QC-mix sample injections should be performed to stabilize the chromatographic system. Furthermore, it allows conditioning the HRMS system, because of a significant performance decrease across the first injections of serum matrix. When QC-mix volume is limited due to scarce volume of study samples, it is also possible to perform the system conditioning by injecting other serum samples; the key aspect is to inject matrix-based samples that are as similar as possible to the serum samples of the study. For example, if the study is conducted on male subjects, it could be possible to use, as conditioning samples, any other male serum samples or a commercially available pool of male serum. The system conditioning is mandatory to obtain reliable results with adequate repeatability.

4. QC-mix sample injections are of crucial importance for the statistical analysis and for the evaluation of the instrument performance in terms of detection and sensitivity of the analytes of interest. The analysis of steroidomic study samples should always start and end with a QC-mix sample injection to enable quality assessment at the beginning and end of the analytical batch; QC-mix injections could also be used during the data treatment for normalization purposes. As a minimum of five QC-mix injections are mandatory to obtain a sufficient robustness for the statistical analysis, these injections should be per-

formed at regular intervals (to be decided taking into account the size of the study, the number of batches and therefore of the length of the analytical sequence) across the sequence between study samples.

5. Study samples are analyzed randomly in blocks of five in minimum between a QC-mix and another in order to reduce the bias of the data. Study sample injections are comprised between two consecutive QC-mix injections, and the size of the block should always be kept the same across all the sequences; the size of the block is resulting from the total number of samples to be analyzed, from the number of QC-mix necessary to have robust statistical results and from the length of the sequence.

6. The end of sample sequence is identical to the beginning, i.e., with two injections of negative serum samples and one of solvent blank. These injections help in evaluating the presence of carry over as well as the increase or decrease of background noise, which can affect the peak integration.

3.8 Data Treatment   Raw data obtained from the UHPLC-HRMS analysis are processed with appropriate software to obtain peak areas of all peaks detected in the steroidomic study samples. Several solutions and commercial and open-source-based strategies could be engaged; however, the following steps should be achieved:

1. The first step of the data treatment is the alignment of the chromatograms. In most of the cases, it is possible to choose between an automatic or manual alignment, in which the user should define the peak of interest that should be used to align the chromatograms. We suggest, in the case of untargeted analyses, where it is not possible to assess the nature of all the peaks in the chromatogram and then decide which are important and which are not, to use the automatic alignment, which uses various peaks as a reference for the alignment.

2. The second and probably one of the most important steps of the data treatment concerns the so-called peak picking. Here, the user has to define where, and how, the software should search for peaks in the aligned chromatograms. It could be appropriate to perform the peak picking only in the retention time window covering the period of the gradient, i.e., to exclude all compounds that are too well retained and elute only in the washing step or those which are not retained at all hence and elute during the dead time. It is also necessary to define a sensitivity threshold (the higher the sensitivity, the larger the number of peaks detected) and the type of adducts. For this step, we suggest using the highest sensitivity in order to detect the maximum number of peaks, which could then be reduced employing statistical tools, and to include only the adducts typical of

steroidal compounds, such as $[M+H]^+$, $[M+H-H_2O]^+$, and $[M+H-2H_2O]^+$. Once the "peak picking" is performed, the software will generate a list of features annotated by a mass (m/z) and a retention time (min); for the analysis of serum sample after SLE extraction, an average of 25,000 features is considered as normal.

3. The last step of the data treatment is represented by the feature annotation. As compound identification still constitutes a major challenge when performing large-scale untargeted analyses for providing biochemical interpretation, it is possible to achieve a different level of annotation (Level 1–3) using both the previously created in-house database and dedicated databases including endogenous and/or exogenous steroids which were developed over the last years, e.g., the Human Metabolite Database (HMDB), LipidMaps, and METLIN.

Firstly, a match with the in-house created database, including the 101 endogenous steroids presented in Table 1, is performed searching among the features generated by the "peak picking." When performing this match, it is mandatory to define the range of uncertainty both for the mass (suggested 5 ppm) and the retention time (suggested 1%). The features which show the same retention time, m/z value, and isotopic pattern compared to a reference endogenous standard present in the database are annotated at Level 1. Then, by working with online databases, it is also possible to predict the retention time of all steroid compounds [14] of the selected database and search if features in the list result in a match with the predicted retention time, m/z value, and isotopic pattern, to achieve a Level 2 annotation. Finally, all features that have a match with the online database but only for m/z and isotopic pattern criteria are assigned with a Level 3 annotation.

When available, the comparison of MS/MS or MS^n fragmentation spectrum with entries of reference experimental MS/MS spectra could also help the annotation process, even if in the case of steroid compounds, which have highly similar fragmentation patterns, the MS/MS information should be preferably used as an exclusion criterion rather than an identification criterion.

---

4   Notes

1. Mobile phase A, as well as weak wash solution, should be prepared fresh every week. This step is important to avoid bacterial growth in aqueous solution used in the chromatographic system.

2. Mobile phase B should be replaced once each 4 months to avoid retention time shifts caused by evaporation and consequent changes in solvent composition.

3. All solvent lines should be primed before starting each analysis, with the aim of eliminating air bubbles that could be formed in the chromatographic system.

4. Chromatographic system pressure should be monitored during the analytical batch and between different batches, to evaluate chromatographic performance. Any significant drift in pressure could influence retention times and therefore data interpretation.

5. If the conducted steroidomic study is composed by more than one analytical batch, it is important to clean the chromatographic column between the batches setting the system to 80% of mobile phase B for a minimum of 30 min. This step helps in reducing the pressure of the system, which can increase after the injection of several extracted serum samples.

6. When setting up the MS method, it could be useful to evaluate in the MS tune software the presence of background noise ions when conditioning the analytical column. Then, if necessary, their m/z values should be added to an exclusion list to avoid the frequent triggering of ddMS2 experiments resulting from these interfering ions.

7. The ESI source should be cleaned before each analytical batch as described in Subheading 3.4. This procedure ensures the best sensitivity and repeatability, hence allowing the comparison between serum samples of a same study injected in different analytical batches.

8. QC samples are of crucial importance for evaluating the performance of the instrument during an analytical sequence. It could be useful to set up an automatic processing method in the Xcalibur environment in order to integrate some target peaks/ compounds and monitor their areas and retention times in QC samples across the whole sequence. Furthermore, when an analytical batch is part of a much larger study, the peak area and the retention time values could also both represent a first approach to compare the performance of the instrument between sequences in different days.

9. Multivariate statistical analysis such as principal component analysis (PCA) is the basic approach to assess the quality of the acquired data when dealing with numerous detected features. For example, it is desirable that in the score plot of PCA, all QC samples are grouped in a unique cluster close to the center of the graph.

References

1. Sanderson JT (2006) The steroid hormone biosynthesis pathway as a target for endocrine-disrupting chemicals. Toxicol Sci 94:3–21

2. Arukwe A (2008) Steroidogenic acute regulatory (StAR) protein and cholesterol side-chain cleavage (P450scc)-regulated steroidogenesis as an organ-specific molecular and cellular target for endocrine disrupting chemicals in fish. Cell Biol Toxicol 24:527–540

3. Shackleton CHL (2012) Role of a disordered steroid metabolome in the elucidation of sterol and steroid biosynthesis. Lipids 47:1–12

4. Kicman AT (2008) Pharmacology of anabolic steroids. Br J Pharmacol 154:502–521

5. World Anti-Doping Agency (WADA), Montreal (2015) Anti-doping testing figures report. http://www.wada-ama.org. Accessed May 2017

6. Basaria S (2010) Androgen abuse in athletes: detection and consequences. J Clin Endocrinol Metab 95:1533–1543

7. Büttner A, Thieme D (2010) Side effects of anabolic androgenic steroids: pathological findings and structure-activity relationships. Handb Exp Pharmacol 195:459–484

8. Fanelli F, Belluomo I, Di Lallo VD et al (2011) Serum steroid profiling by isotopic dilution–liquid chromatography–mass spectrometry: comparison with current immunoassays and reference intervals in healthy adults. Steroids 76:244–253

9. Rosner W, Auchus RJ, Azziz R et al (2007) Position statement: utility, limitations, and pitfalls in measuring testosterone: an endocrine society position statement. J Clin Endocrinol Metab 92:405–413

10. Handelsman DJ, Wartofsky L (2013) Requirement for mass spectrometry sex steroid assays in the journal of clinical endocrinology and metabolism. J Clin Endocrinol Metab 98:3971–3973

11. Wartofsky L, Handelsman DJ (2010) Standardization of hormonal assays for the 21st century. J Clin Endocrinol Metab 95:5141–5143

12. World Anti-Doping Agency (WADA), Montreal (2016) endogenous anabolic androgenic steroids, measurement and reporting, technical document TD2016EAAS. http://www.wada-ama.org. Accessed May 2017

13. Sjovall J (2004) Fifty years with bile acids and steroids in health and disease. Lipids 39:703–722

14. Randazzo GM, Tonoli D, Strajhar P et al (2017) Enhanced metabolite annotation via dynamic retention time prediction: steroidogenesis alterations as a case study. J Chromatogr B Analyt Technol Biomed Life Sci. 1071:11–18

Chapter 19

Metabolomics in Human Acute-Exercise Trials: Study Design and Preparation

Aikaterina Siopi and Vassilis Mougios

Abstract

Metabolomics can be of great value in the study of exercise metabolism. However, because of the high intraindividual and interindividual biological variability of the human metabolome, special considerations should be taken into account when designing an acute-exercise metabolomic study. To study different exercise parameters, e.g., different exercise modes, intensities, etc., a crossover study design, where each participant acts as their own control, is preferable to a parallel design, one involving different groups of participants. Moreover, the study should include a no exercise, control trial. Before each trial, participants should follow carefully designed preparatory steps to control for possible confounding factors, i.e., maintain repeatable and constant conditions for all individual trials of the study to minimize variation due to factors other than the one(s) being studied. This chapter focuses on the design of human metabolomic studies, where the intervention is an acute metabolic challenge, such as an exercise bout or a test meal, and presents some basic steps for screening potential participants, performing preliminary tests, preparing for the trial day, and performing the trial.

Key words Metabolomics, Acute exercise, Study design, Humans

1 Introduction

Metabolomics can be of great value in the study of exercise metabolism. It can help to decipher the molecular mechanisms behind exercise-induced metabolic responses and benefits. The response of the metabolome to exercise may be indicative of an individual's fitness or disease status; therefore, exercise metabolomics could be used for diagnostic/prognostic purposes [1]. Moreover, metabolomics can prove very useful in assessing the effectiveness of exercise interventions and training regimens. Ultimately, all these could translate into personalized exercise prescription on the basis of an individual's metabolic profile at rest or in response to exercise. Therefore, metabolomics have been increasingly applied in exercise intervention studies during the past decade [1–7]. Currently,

Georgios A. Theodoridis et al. (eds.), *Metabolic Profiling: Methods and Protocols*, Methods in Molecular Biology, vol. 1738, https://doi.org/10.1007/978-1-4939-7643-0_19, © Springer Science+Business Media, LLC, part of Springer Nature 2018

research efforts are focusing on recording a comprehensive chart of the molecular changes caused by exercise.

When the aim is to study different exercise parameters, e.g., different exercise modes or intensities, a crossover design, where each participant serves as their own control, seems to be preferable to a parallel design, which includes different groups of participants [8]. The reason is the very high interindividual variability of the human metabolome. Let us assume that the study objective is to compare the effects of three different exercise types on the metabolome. In a parallel study design, we would randomize the participants to three groups, and each group would perform exercise of one type. In a crossover design, however, all of the participants would perform all three exercises in a randomized order. The latter design implies more visits for each participant, but it has the advantage that their metabolic responses to one exercise type will be compared to their own responses to another exercise type, not the responses of other individuals. However, if the study objective is the effect of a specific health condition or disease on the response of the metabolome to exercise, a parallel design, with a carefully matched control group, is inevitable.

In an acute-exercise metabolomic study, it is also necessary to include a no exercise trial. This should be identical to the exercise trials in all aspects except that the participants will rest instead of exercising. At the end, the post-exercise metabolic profiles or fingerprints will be compared to that of the resting trial in order to isolate the effects of exercise from other confounding factors such as diurnal variation or fasting. Moreover, the samples from the resting trial can be used to correct for the batch effect, which hinders metabolomic analysis, especially when based on LC-MS platforms. As long as all samples from one participant are analyzed in the same batch, normalizing the data for each metabolite in the samples of the exercise trials to the respective value in the resting trial could correct for unwanted batch effects [9].

Regarding the choice of biological matrix, blood plasma/serum and urine are biofluids that have been traditionally used to monitor health status, since they are easily collected and reflect the global state of an individual. Urine may have an advantage over blood for biomarker identification purposes, as it is under no homeostatic mechanisms. Change is the most important aspect of biomarker discovery. Metabolite changes in blood could actually be detected in urine with higher sensitivity, since blood changes tend to be quickly normalized by strict homeostatic mechanisms. Urine is accessible noninvasively, so it is easier to collect in large quantities and through multiple time points to study time-dependent changes in metabolites. Moreover, urine is more stable and less complex than other biofluids, therefore, an ideal source of biomarkers [10, 11].

In the following sections, we will present all the recommended materials and procedures when preparing for an acute-exercise metabolomic study in humans.

2 Materials

The following instruments are recommended:

- Sphygmomanometer
- Automatic biochemistry analyzer for preliminary biochemical screening.
- A device for body composition analysis such as skin calipers, bioelectrical impendence analysis (BIA), or dual-energy X-ray absorptiometry (DXA).
- Indirect calorimetry metabolic cart—oxygen and carbon dioxide analyzers.
- Treadmill or cycle ergometer.
- Heart rate monitor.
- Pedometer (step counter).
- Nutritional analysis software.

3 Methods

3.1 Screening Potential Participants

The screening process is necessary to determine whether an individual meets the criteria to participate in the study. There should be clearly defined inclusion and exclusion criteria. The inclusion criteria depend on the particular target population of the study, i.e., gender, age, lifestyle (e.g., sedentary or active), training status, health status, etc. The exclusion criteria usually aim to eliminate possible confounding factors and to decrease biological variation, which is a big problem in metabolomic studies. Exclusion criteria often include the presence of acute or chronic disease (other than the one perhaps being tested), any contraindication for exercising, use of medication or supplements, smoking, and dieting or recent change in body weight (e.g., >2 kg within 6 months). The usual steps for the screening process are:

1. Prescreening/phone assessment. Using a form that you will have created for this purpose, record the individual's personal information, including contact information, information regarding all inclusion and exclusion criteria, any previous participation in other research studies, as well as availability (days and hours) for participating in the study (*see* **Note 1**).

2. First visit. If a volunteer is eligible to continue, arrange a first appointment for them to visit the research facilities and meet the staff. During the meeting, explain all steps of the study thoroughly (*see* **Note 2**), and have the volunteer read and sign the necessary consent forms, while you answer any questions they may have (*see* **Note 3**). Help the volunteer complete all necessary paperwork, such as medical history, physical activity questionnaires, and dietary questionnaires.

3. Medical examination and screening measurements. (This step may or may not be carried out during the first visit, depending on what is convenient.) Collect basic anthropometric data (e.g., height, weight, and waist circumference), vital signs (e.g., blood pressure and heart rate), and a blood sample for biochemical screening (under fasted conditions). Medical clearance for participating in exercise trials should follow the established guidelines [12]. More than one visit may be necessary to complete the screening process.

3.2 Preliminary Tests

After the screening process and before the trials, participants usually need to go through some preliminary tests (*see* **Note 4**). These can include:

1. Body composition measurements, such as lean body mass, total fat mass, and visceral fat mass (*see* **Note 5**). Make sure the participants have followed the necessary preparations for the respective analysis. These measurements can be used as descriptive characteristics of the study sample. Moreover, they can be correlated with outcomes of the metabolomic analysis.

2. Measurement or estimation of resting energy expenditure (*see* **Note 6**). This measurement will be used to design the dietary plan of the participants during the study in order to decrease intraindividual and interindividual variations.

3. Cardiorespiratory fitness assessment. Perform a maximal or submaximal incremental test (depending on your study sample) to assess maximal oxygen consumption (VO_2max) and maximal heart rate [12]. This measurement may be used to set the exercise intensity for endurance exercise trials.

4. Muscular strength measurement. Perform a determination or prediction (depending on your study sample) of one-repetition maximum [13]. This measurement may be used to set the exercise intensity for resistance exercise trials.

3.3 Preparation for the Exercise Trials

Arrange some extra time at the end of the preliminary tests appointment to talk with each participant about the scheduling and preparation for the trials. Again, be thorough and provide all instructions in writing as well as verbally for the participants to take home (*see* **Note 2**). Schedule the next appointment (*see* **Notes 4** and **7**). A

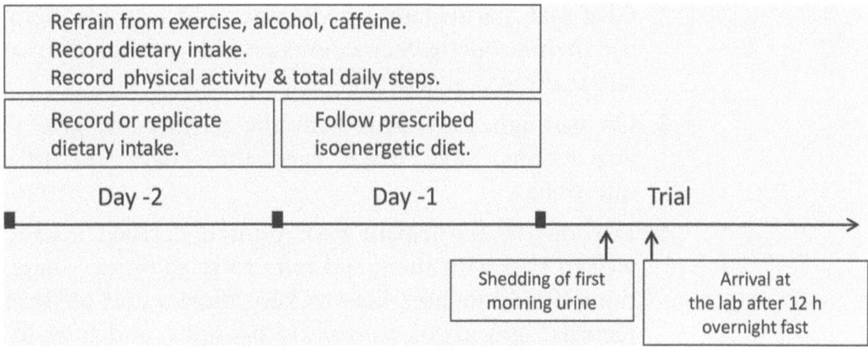

Fig. 1 Overview of the preparation for each trial of the study (*see* **Note 10**)

usual practice is to space the trials 1 week apart. Do not exceed 2 weeks to avoid the risk of losing comparability among trials. For premenopausal women, each trial should be scheduled at the same phase of the menstrual cycle [14].

The preparation for each trial usually begins 2 days before (Fig. 1). The aim is to control for possible confounding factors. You have to maintain repeatable and constant conditions for all individual trials of the study to minimize biological variation due to factors other than the one being studied (i.e., exercise). This step is critical for the success of an acute human metabolomic study, where variation is a huge issue. Take the time to help participants comply with the instructions (*see* **Note 8**).

Participants need to record their diet [15] *and* physical activity for the agreed period (1 or 2 days) before the first trial in order to replicate it before the next trial(s), thus minimizing intraindividual variance. To minimize interindividual variance, all participants should have equivalent dietary intakes: they should follow isoenergetic diets (i.e., diets matching energy intake to energy expenditure) or, in case a dietary intervention is planned together with the exercise intervention, cause the same energy deficit or surplus to all participants. To this end, a qualified dietitian should design personalized dietary plans for at least the day before the metabolomic trial (*see* **Note 9**). Participants should follow their plans but nevertheless record what they actually ate. At the end of the study, all dietary records must be analyzed to validate that there was no significant nutritional difference that could have affected the results of the metabolomic analysis. Unless otherwise dictated by the design of the study, participants should be instructed to refrain from intense exercise, caffeine intake, and alcohol consumption on the day before the trial, as all these may have a reverberation on metabolism.

To prepare for the first trial:

1. Provide each participant with special forms to record their daily dietary intake, physical activity, and step count. They usually need to start recording all these 2 days before the trial (*see* **Note 10**).

2. Give each participant a pedometer and instruct them how to
   use them properly. Step counts provide a rough estimate of the
   habitual physical activity.

3. Go through the forms with the participant, give thorough
   instructions, and make sure they have no unanswered
   questions.

4. Provide the participant with printed detailed instructions as
   well to take with them and refer to at all times. These instructions should include how to keep dietary and physical activity
   records, how to measure food portions, and how to use the
   pedometer. Remind the participant to avoid intense exercise,
   caffeine intake, and alcohol consumption.

5. Provide printed detailed instructions for the night before and
   the morning of the trial day. These instructions usually include
   having the last meal 12 h before their morning appointment in
   the lab (but no more than 14 h), avoiding any food consumption at home in the morning, shedding of first morning urine if
   there will be urine sampling, etc.

6. Provide each participant with contact information of a member
   of the research team to refer to for any questions or problems
   that may arise during the preparation for the trials.

7. Ask them to bring all records with them at all visits.

3.4 Trial Day

After arrival of the participant to the lab and before starting the
trial:

1. Weigh them to make sure that there has been no considerable
   change since the previous measurement.

2. Go with them through a checklist to make sure they have successfully followed all necessary steps of the preparation
   process.

3. Check their dietary and physical activity records of the preced-
   ing days.

4. Take a 24-h dietary recall to cross-validate the dietary record of
   the previous day and to make sure it has been completed properly (*see* **Note 11**).

5. If they have failed to follow any of the preparation steps and you
   feel that the reliability of the process has been compromised,
   you may have to reschedule the trial.

   During the trial:

1. Have the participants remain seated and relaxed during the
   non-exercising parts of the trial.

2. Mark their water consumption during the first trial and have
   them replicate it during the next trial(s).

3. Maintain constant environmental conditions (room temperature and humidity).

4. Maintain a stress-free environment as possible (the room should be tidy, quiet, as private as possible, not crowded, etc.).

At the end of the first trial, schedule the next trial and go through the preparatory instructions with the participants (*see* **Note 12**). For all trials participants have to follow exactly the same preparatory steps as they did for the first trial (Fig. 1). If a participant's dietary intake for the day before the trial is even slightly different from the prescribed dietary plan, they should repeat their actual dietary intake (not the prescribed) as recorded in the respective dietary record (and cross-validated with the 24-h dietary recall).

4 Notes

1. Even if there is compensation for the participants and a large number of candidates, usually the necessary strict criteria make recruitment difficult. Only a small portion of those initially interviewed will continue through the screening process.

2. Be thorough when explaining the study process to the participants. Provide written information/instructions as well. If possible, use figures and/or tables to give an overview of the study. Nonetheless, remember to repeat the process/instructions at every step of the way to ensure maximal compliance. It is your job to guide the participants throughout the process and help them comply with the instructions. Compliance is very important to ensure comparability among study trials or groups and to control for the high intraindividual and interindividual biological variance that complicates metabolic profiling.

3. All procedures have to be approved by the institutional review board and comply with the Helsinki declaration of 1975, as revised in 2013.

4. After enrollment of a participant to the study, it is important that they complete all trials as soon as possible. Any unnecessary delays increase the chances for dropping out, losing eligibility to participate, or losing comparability of the trials.

5. The gold standard for body composition analysis is DXA. However, this method requires expensive and non-portable equipment. A practical compromise can be the method of BIA. Preparation for this analysis includes abstention from drinking water or other liquids for the last 4 h; abstention from eating for the last 12 h; abstention from exercising, drinking caffeine, drinking alcohol, using a sauna, or using a

hot tub for the last 24 h; and abstention from diuretics for the last 7 days prior to the analysis.

6. If possible, use the method of indirect calorimetry with the canopy technique to measure resting energy expenditure. Alternatively, you can use equations to estimate energy needs [16].

7. In a crossover design, each participant must complete all trials in a randomized order. You can use a random-number generator software (one of the many available online for free) to obtain random sequences.

8. In human trials, the interpersonal communication skills of the researcher/staff are very important.

9. The dietary plan should be carefully designed to maximize compliance and avoid major effects of diet on the metabolome. It should not include complicated recipes, expensive ingredients, or food that the participant does not like, is not used to, or is allergic to. Ideally, the research project should provide catering for all meals of the participants.

10. Depending on how many sampling points there are throughout the day of each trial, you may need to control diet and physical activity on the day of the trial as well.

11. Dietary recalls need to be taken by a trained and experienced professional to be accurate.

12. The best way to match acute-exercise bouts is to measure exercise energy expenditure by indirect calorimetry.

References

1. Lewis GD, Farrell L, Wood MJ et al (2010) Metabolic signatures of exercise in human plasma. Sci Transl Med 2(33):33ra37. https://doi.org/10.1126/scitranslmed.3001006

2. Daskalaki E, Easton C, Watson DG (2014) The application of metabolomic profiling to the effects of physical activity. Curr Metabol 2(4):233–263

3. Pechlivanis A, Kostidis S, Saraslanidis P et al (2013) 1H NMR study on the short- and long-term impact of two training programs of sprint running on the metabolic fingerprint of human serum. J Proteome Res 12(1):470–480. https://doi.org/10.1021/pr300846x

4. Pechlivanis A, Papaioannou KG, Tsalis G et al (2015) Monitoring the response of the human urinary Metabolome to brief maximal exercise by a combination of RP-UPLC-MS and (1)H NMR spectroscopy. J Proteome Res 14(11):4610–4622. https://doi.org/10.1021/acs.jproteome.5b00470

5. Enea C, Seguin F, Petitpas-Mulliez J et al (2010) (1)H NMR-based metabolomics approach for exploring urinary metabolome modifications after acute and chronic physical exercise. Anal Bioanal Chem 396(3):1167–1176. https://doi.org/10.1007/s00216-009-3289-4

6. Nieman DC, Gillitt ND, Sha W (2015) Metabolomics-based analysis of banana and pear ingestion on exercise performance and recovery. J Proteome Res 14(12):5367–5377. https://doi.org/10.1021/acs.jproteome.5b00909

7. Peake JM, Tan SJ, Markworth JF (2014) Metabolic and hormonal responses to isoenergetic high-intensity interval exercise and continuous moderate-intensity exercise. Am J Physiol Endocrinol Metab 307(7):E539–E552. https://doi.org/10.1152/ajpendo.00276.2014

8. Heinzmann SS, Merrifield CA, Rezzi S et al (2012) Stability and robustness of human met-

abolic phenotypes in response to sequential food challenges. J Proteome Res 11(2):643–655. https://doi.org/10.1021/pr2005764

9. Siopi A, Deda O, Manou V (2017) Effects of different exercise modes on the urinary metabolic fingerprint of men with and without metabolic syndrome. Metabolites 7(1). https://doi.org/10.3390/metabo7010005

10. Wu J, Gao Y (2015) Physiological conditions can be reflected in human urine proteome and metabolome. Expert Rev Proteomics 12(6):623–636. https://doi.org/10.1586/14789450.2015.1094380

11. Li M (2015) Urine reflection of changes in blood. Adv Exp Med Biol 845:13–19. https://doi.org/10.1007/978-94-017-9523-4_2

12. American College of Sports Medicine (2016) ACSM's guidelines for exercise testing and prescription, 10th edn. Wolters Kluwer, Philadelphia, PA

13. Dohoney P, Chromiak JA, Lemire D et al (2002) Prediction of one repetition maximum (1-RM) strength from a 4-6 RM and a 7-8 RM submaximal strength test in healthy young adult males. J Exercise Physiol 5:54–59

14. Wallace M, Hashim YZ, Wingfield M et al (2010) Effects of menstrual cycle phase on metabolomic profiles in premenopausal women. Hum Reprod 25(4):949–956. https://doi.org/10.1093/humrep/deq011

15. Walsh MC, Brennan L, Malthouse JP (2006) Effect of acute dietary standardization on the urinary, plasma, and salivary metabolomic profiles of healthy humans. American. Am J Clin Nutr 84(3):531–539

16. Harris JA, Benedict FG (1918) A biometric study of human basal metabolism. Proc Natl Acad Sci U S A 4(12):370–373

INDEX

Georgios A. Theodoridis et al. (eds.), *Metabolic Profiling: Methods and Protocols*, Methods in Molecular Biology, vol. 1738,
https://doi.org/10.1007/978-1-4939-7643-0, © Springer Science+Business Media, LLC, part of Springer Nature 2018

MIX
Papier aus verantwortungsvollen Quellen
Paper from responsible sources
FSC® C105338

If you have any concerns about our products,
you can contact us on
ProductSafety@springernature.com

In case Publisher is established outside the EU,
the EU authorized representative is:
**Springer Nature Customer Service Center GmbH
Europaplatz 3, 69115 Heidelberg, Germany**

Printed by Libri Plureos GmbH
in Hamburg, Germany